# Flying Aircraft Carriers of the USAF: Wing Tip Coupling

Brian Lockett

LockettBooks
www.Air-and-Space.com

**Acknowledgements:**

Louie Alley (Air Force Safety Center), Clarence E. "Bud" Anderson, Archie DiFante (Air Force Historical Research Agency), Beryl Erickson, Bruce Hess (Aeronautical Systems Center History Office), Leonard Hillary, Meyers Jacobson, Dennis Jenkins, Fred Johnsen (Air Force Flight Test Center History Office), Frank Kleinwechter, George Krieg, John McDowell, General Dynamics, Dave Menard (US Air Force Museum), Michael Moore (Lockheed-Martin Tactical Aircraft Systems), Ray Puffer (Air Force Flight Test Center History Office), Jim Solenberger, Greg Spahr, Roy Wendell (Fairchild Republic Corporation), and Harold Wood

**Dedicated to my father,**
**Richard Lockett,**
**who inspired my interest in all things aeronautical.**

**About the author:**

Brian Lockett earned his Private Pilot certificate as a teenager. He has been a member of the American Aviation Historical Society since 1973, and the Journal of the A.A.H.S. has published seven of his aviation history articles. He graduated from the University of California at Santa Barbara with a Bachelor of Arts degree in Geography. Brian worked for nearly a decade at the Santa Barbara County Air Pollution Control District. He is the creator of the Goleta Air & Space Museum web site (www.Air-and-Space.com). His username is GoletaBrian on the Gigapan, Panoramio, Weather, and YouTube web sites. He publishes calendars, DVDs, and books as LockettBooks at Lulu.com.

**Also by Brian Lockett:**

*Flying Aircraft Carriers of the USAF: Project FICON*
*Balls Eight: History of the Boeing NB-52B Stratofortress Mothership*
*Painting Guide for the Boeing Stratofortress Motherships*

LockettBooks
www.Air-and-Space.com


First Edition

ISBN 978-0-578-03186-6

# Contents

*Douglas C-47A-35-DL Skytrain 42-23918 and Culver PQ-14B-CL 44-68334. (Bud Anderson)*

## Section 1: C-47/Q-14 Wing Tip Coupling

### *Concept development*

During and after World War II, aircraft designers were faced with the problem of increasing the range of strategic bombers. During the war, military planners were faced with the possibility that air bases in England would become unavailable to U. S. bombers. Strikes against the axis powers in Europe might have to be conducted from bases in North America. In the early years of the cold war, the vast size of the Soviet Union taxed the range capabilities of U.S. warplanes.

The most obvious way to increase the range of an airplane is to load more fuel on it. To carry more fuel without increasing the wing loading, the wing must be made larger. It is not practical to simply make the wingspan longer. The structure of the wing root has to be redesigned to support the extra lift that the larger wing generates.

Dr. Richard Vogt, a German immigrant to the United States, proposed that floating wing panels carrying fuel tanks could be attached to the wing tips of an airplane with hinges to extend its range without requiring a redesign of the wing root. The floating wing panels would support their own weight, without increasing the load on the airplane's wings.

The fuel carrying wing extensions would increase the wingspan of the airplane, effectively increasing the aspect ratio.[1] Increasing the aspect ratio of a wing improves its aerodynamic efficiency. A long, slender wing loses less energy to induced drag than a wide wing of short span that has the same area. It is for this reason that sailplanes are equipped with such long and slender wings.

---

1 The ratio of the span of a wing divided by the wing chord is referred to as the aspect ratio of the wing.

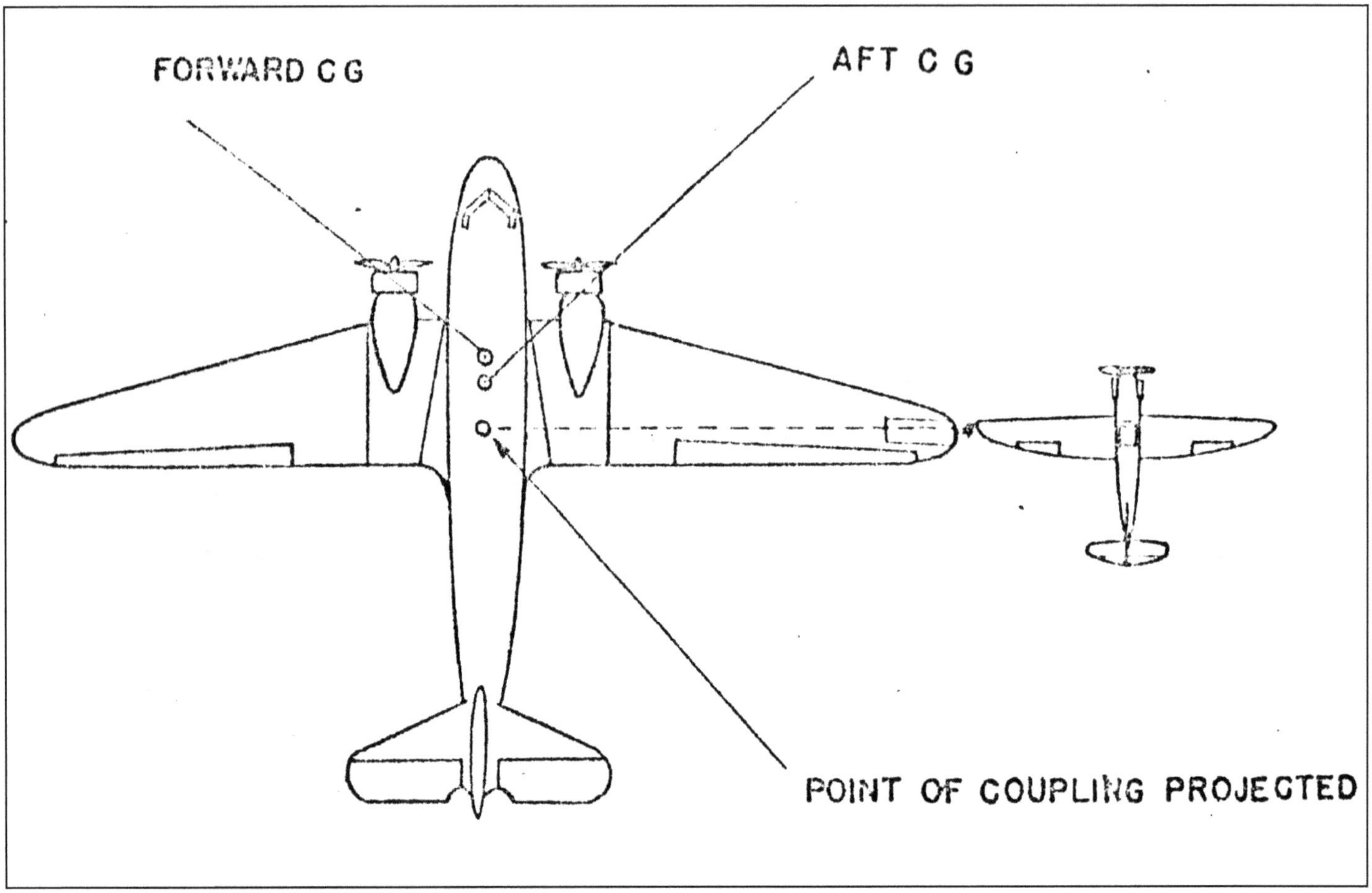


*Plan view of the coupled C-47 and Q-14. (Bud Anderson)*

The improved efficiency of a wing with self-supporting fuel tanks would offset a large fraction of the drag of the tanks. It was predicted that extra fuel could be carried with a relatively small increase in the overall drag of the bomber.

The Air Force initiated a project to simulate floating wing panels with a piloted light plane that coupled to a larger airplane in flight. Soon the scope of the project expanded to explore the possibility of towing fighters coupled to the wing tips of bombers.

### Antecedents

Dr. Vogt originated the concept of extending the range of airplanes by connecting them at the wing tips during World War II. The German Research Institute for Gliders (DFS) conducted a preliminary evaluation of his idea in 1945. Two light airplanes of similar type were equipped with coupling mechanisms on opposite wing tips. Their coupling mechanism was a ball joint that allowed the, aircraft to roll, pitch, and yaw with respect to one another.

*Douglas C-47A-35-DL Skytrain 42-23918 and Culver PQ-14B-CL 44-68334 on the ramp at the Wright Air Development Center. (Air Force Film 14425: Investigation of Aircraft in Coupled flight)*

*Culver PQ-14B-CL 44-68334. Orthochromatic film makes the Crimson Q-14 appear dark in photos taken from the C-47. (Bud Anderson)*

The aircraft were connected with a tether before take-off. They took off in formation with the tether hanging slack between them. Once airborne, the pilot in one of the airplanes activated a winch to pull the two airplanes together. The pilots maintained formation flight while the airplanes were pulled together. Once coupled, they evaluated the behavior of the coupled airplanes in turns and climbing and descending flight.

The project was only intended to evaluate the handling qualities of the coupled airplanes. No attempt was made to evaluate the techniques required to couple two airplanes that had taken off separately.

Following the war, Dr. Vogt immigrated to the United States under the auspices of operation Paperclip." He promoted his value to the United States by presenting the concept for self-supporting, winged fuel tanks to researchers at the Wright Air Development Center at Wright-Patterson Air Force Base, Ohio.

It was felt that the easiest means of making a preliminary investigation of the idea would be to simulate the free-floating fuel tank with a small manned airplane that hooked on to a larger test airplane in flight. The flight test concept immediately inspired the idea that manned fighters could attach themselves to the wing tips of a bomber to be towed to the target area with only a small reduction in the range of the bomber.

The Boeing B-19 *Superfortress* was an obvious choice for the basis of a composite airplane design. The performance of the B-29 far exceeded that of any other strategic bomber of World War II and the Air Force had many *Superfortresses* in operation. The Republic F-84 *Thunderjet* was considered to be the most suitable jet fighter design for the towed fighter role.

### *Aircraft Selection*

The 2750th Air Base Group at Wright-Patterson Air Force Base conducted an initial investigation of wing tip coupling in the summer of 1949. Very little was known about the methods and hazards of coupling airplanes together in flight. Although the full-scale towed fighter project envisioned the use of a B29 *Superfortress* bomber and a pair of F-84 *Thunderjet* fighters, it was considered expedient to conduct a preliminary evaluation with smaller airplanes using a coupling mechanism of relatively simple design.

The aircraft types selected for initial testing were a Douglas C-47A and a Culver Q-14B. They were chosen for the project partly because they had mass and wingspan ratios that were similar to those of the B-29 and F-84. The (C-47 had 12 times the mass of the Q-14B while the B-29 weighed ten times what the F-84 weighed. The C-47 wingspan was 3.2 times as long the wingspan of the Q-14B. The B-29 wingspan was 3.6 times as long as the F-84 wingspan.

The Douglas C-47 was famous as the military transport version of the DC-3. The wingspan of the C-47 was 95 feet and the length was almost 64 feet. The C-47 was equipped with a pair of 1,200 horsepower Pratt and Whitney R-1830 radial engines. The gross weight of the C-47 was nearly 25,000 pounds.

The Culver Q-14B was a target drone that was equipped with a cockpit for ferry flights. The design was based on the Culver *Cadet* general aviation airplane. It had a wingspan of 30 feet and was 19 and 1/2 feet long. The Q-14B was fitted with a Franklin 0-300 four-cylinder, horizontally opposed engine.

There were many C-47s available following the war. It was not difficult to find one to assign to the project. C-47A-35-13-DI 42-23918 was already based at Wright-Patterson Air Force Base, so it was selected for the wing tip coupling tests.

C-47A 42-23918 was manufactured by the Douglas Aircraft accompany at Long Beach, California. It carried Douglas construction number 9780. It was delivered to the Army Air Corps on June 26, 1943. Its initial assignment was Alliance Field, Nebraska. In August 1943 it was assigned to Sedalia Field in Missouri. In May 1945 it was transferred to Troop Carrier Command and assigned to the 316th Troop Carrier Group at Pope Field, North Carolina. It was transferred to the Tactical Air Command in March 1946 and assigned to the 439th Troop Carrier Group at Sedalia Field in Missouri: It was reassigned to the 434th Troop Carrier Group at Greenville Field, South Carolina, the following June.

At the time of the formation of the United States Air Force, 42-23918 was transferred to the Air Material Command. It was assigned to the 4000th Base Unit at Wright-Patterson Air Force Base, Ohio, in August 1947. Three months later it was reassigned to the 2750th Air Base Wing, which was also based at Wright-Patterson Air Force Base. It was re-designated EC47A in recognition of its specialized testing function.

PQ-14B 44-68334 was manufactured by Culver Aircraft Company in Wichita, Kansas. The Army Air Corps took possession of 44-68334 on March 13, 1945. It was delivered to Jackson Field, Mississippi, and immediately placed into storage for the rest of the year. 44-68334 was transferred to the Air Technical Service Command in January 1946. It was assigned to the 4136th Air Base Unit at Tinker Air Force Base, Oklahoma, and placed into storage again.

At the time that the drones were procured by the Army Air Corps, they were designated PQ-14. "PQ" was the Army's designation prefix for manned target drones. Later, the Air Force used the designation "Q" for aerial targets.

When the United States Air Force was created in 1947, 44-68334 was re-designated Q-l4B and transferred to the Air Materiel Command. It was assigned to the 4145th Air Base Unit at Holloman Air Force Base, New Mexico, for nearly two years.

Q-14B 44-68334 arrived at Wright-Patterson Air Force Base in March 1949. It was assigned to the 2750th Air Base Group, which assigned it to the wing tip coupling investigation.

### *Modifications*

The Aircraft Laboratory at Wright-Patterson Air Force Base developed a mechanism to couple the Douglas EC-47A and Culver Q-14B together in flight. The coupling mechanism was relatively simple in comparison to that envisioned for the full-scale project. The coupling design for the full-scale project fixed the F-84 to the B-29 wing in pitch and yaw. The coupling would only allow the F-84 to roll about the wing tip of the B-29.

For the sake of simplicity, the coupling mechanism for the preliminary investigation was a ball joint attachment that allowed the Q-14B to roll, pitch, and yaw relative to the EC47A The ball joint allowed nearly unlimited freedom in roll. The Q-14B was limited to 22 degrees of rotation in pitch and 22 degrees in yaw.

### *EC-47A*

The Experimental Fabrication Laboratory constructed the wing tip coupling mechanism. A standard 42-G-8290, glider tow release mechanism was mounted on the upper surface of the right wing of the EC-47A. It had a towing capacity of 9,500 pounds. The release mechanism could be used to jettison the Q-14B if it posed a hazard to the EC-47A.

The wing tip of the EC-47A was substantially reinforced to support the tow release mechanism. Two l/4-inch thick duralumin plates were bolted to the top and bottom of the wing for reinforcement. The plates were 15 inches by 12.5 inches wide and were attached with twenty-four S|/l6-inch diameter aircraft-steel bolts that passed through the wing. Spacers were placed around the bolts inside the wing to maintain the proper contours of the skin. l/8-inch thick, rubber gasket sheets were placed between the plates and the skin of the wing to ensure a tight fit. S|/8-inch diameter, aircraft-steel bolts were used to attach the tow release to the wing.

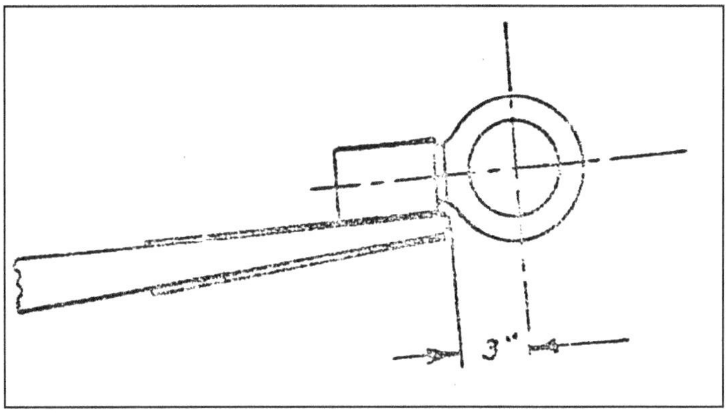


*Coupling ring installation on the C-47 wing tip. The center of the coupling ring is three inches from the wing tip. (Bud Anderson)*

A steel ring with a clinch inside diameter was welded to the d-ring of the tow release to serve as a receptacle for a lance mounted on the wing tip of the Q-14B. The ring was mounted three inches from to: wing tip of the EC-47A. The cable for the tow release mechanism at the wing tip was connected to the standard tow release cable running to the tail of the EC-47A. The pilot of the C-47 used the standard tow release handle in the cockpit to jettison the Q-14B.

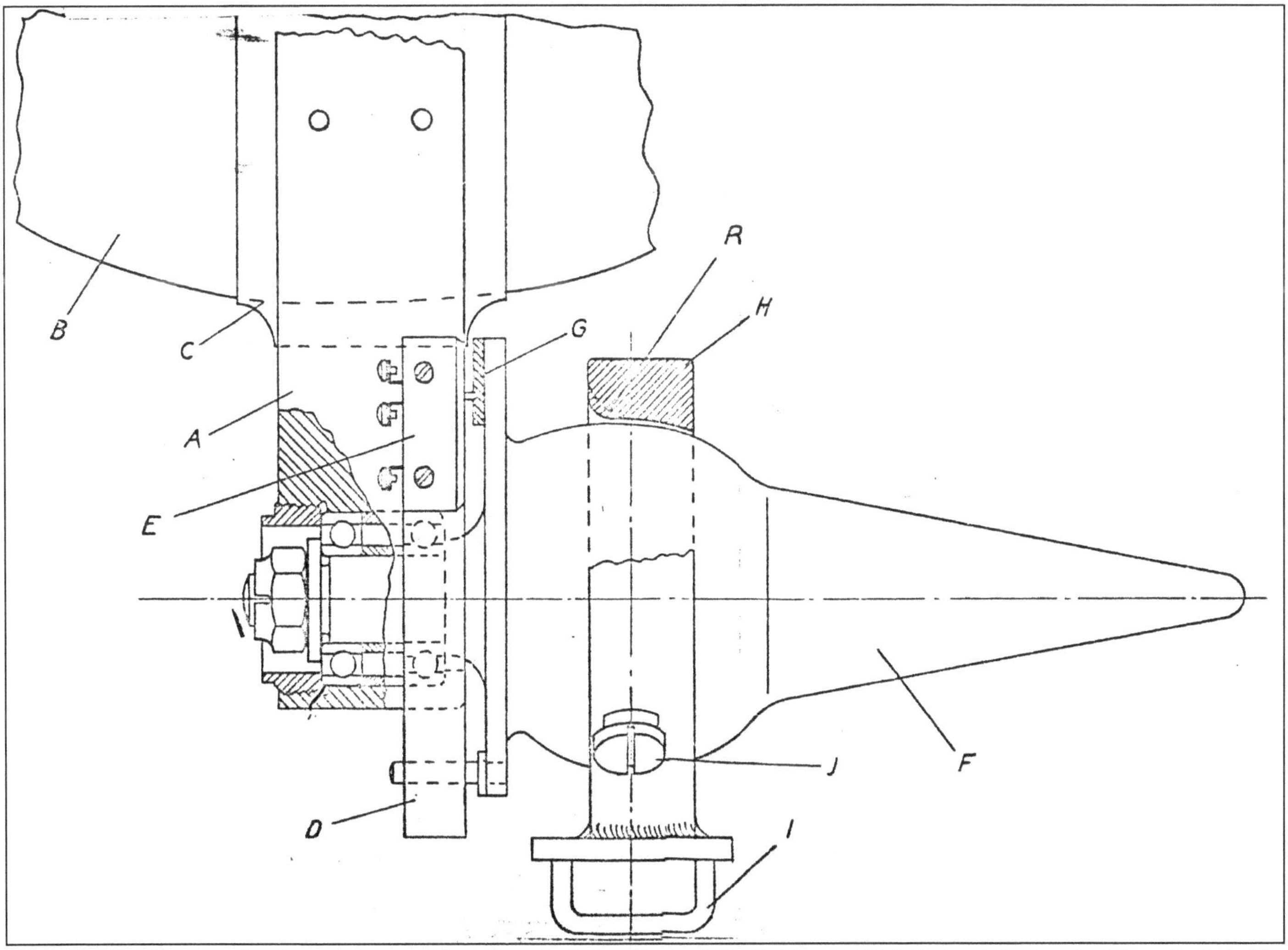


*Wing tip coupling mechanism: A. Lance Mounting, B. Wing Tip Q-14B, C. Wood Fillet, D. Leaf Spring, E. Micro Switch, F. Coupling Lance, G. Micro Switvh Actuator, H. Couplilng Ring, D-Ring, I. Engaging Doc (Bud Anderson)*

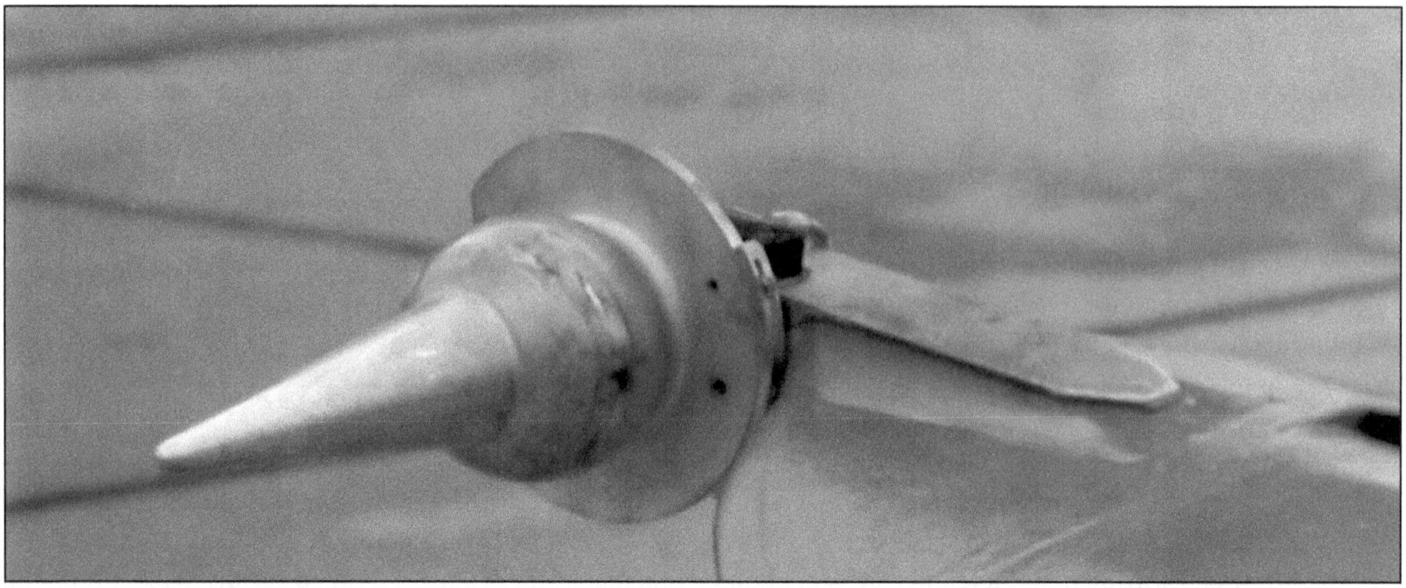

*The coupling lance on the wing tip of Culver PQ-14B-CL 44-68334 points to the rear. (Air Force Film 14425)*

*The coupling ring fit snugly wround the spherical base of the lance. It was free to rotate in pitch, yaw, and roll. (Air Force Film 14425)*

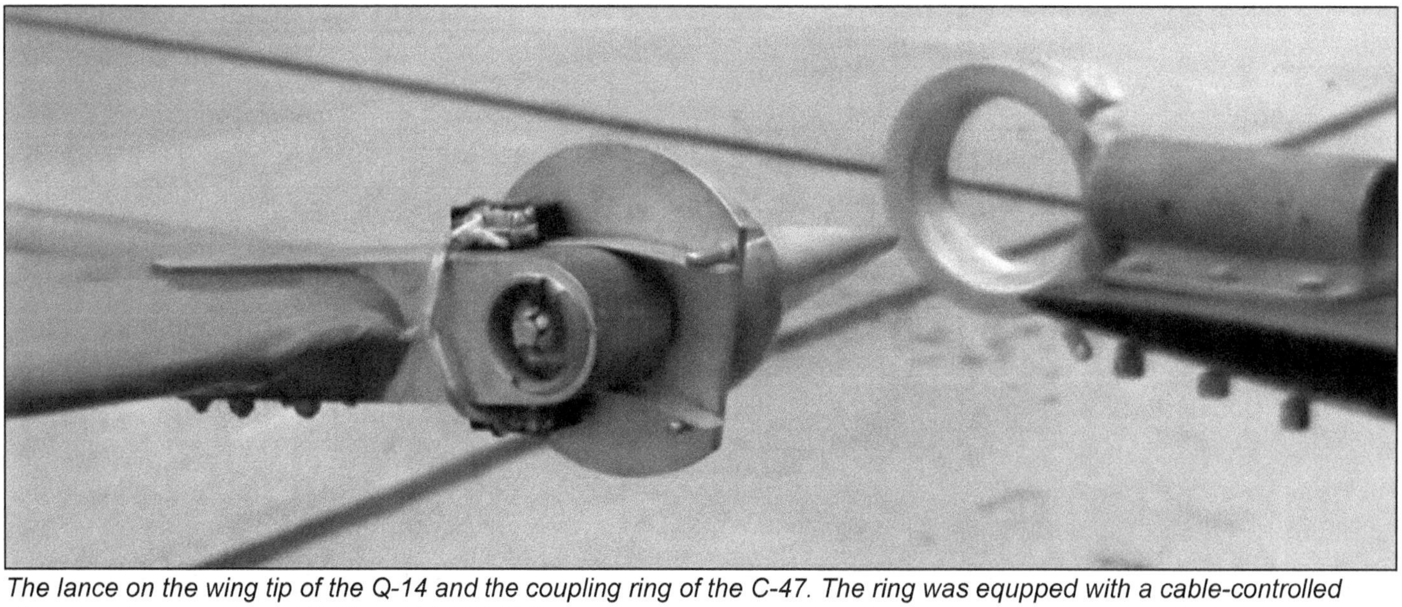

*The lance on the wing tip of the Q-14 and the coupling ring of the C-47. The ring was equpped with a cable-controlled disconnect mechanism.. (Air Force Film 14425)*

degrees of rotation in pitch and 22 degrees in yaw.

### *EC-47A*

The Experimental Fabrication Laboratory constructed the wing tip coupling mechanism. A standard 42-G-8290, glider tow release mechanism was mounted on the upper surface of the right wing of the EC-47A. It had a towing capacity of 9,500 pounds. The release mechanism could be used to jettison the Q-14B if it posed a hazard to the EC-47A.

The wing tip of the EC-47A was substantially reinforced to support the tow release mechanism. Two l/4-inch thick duralumin plates were bolted to the top and bottom of the wing for reinforcement. The plates were 15 inches by 12.5 inches wide and were attached with twenty-four S|/l6-inch diameter aircraft-steel bolts that passed through the wing. Spacers were placed around the bolts inside the wing to maintain the proper contours of the skin. l/8-inch thick, rubber gasket sheets were placed between the plates and the skin of the wing to ensure a tight fit. S|/8-inch diameter, aircraft-steel bolts were used to attach the tow release to the wing.

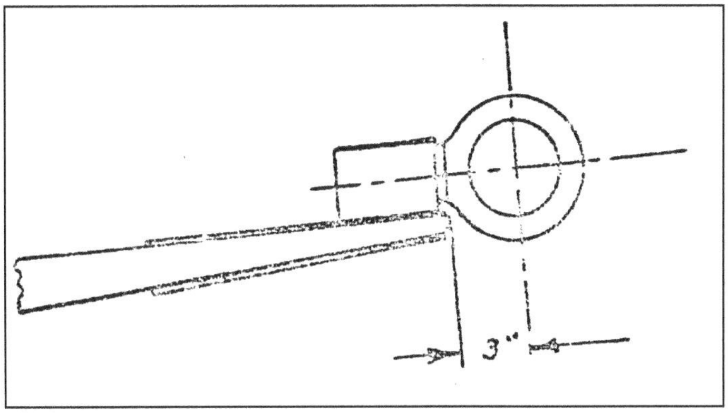


*Coupling ring installation on the C-47 wing tip. The center of the coupling ring is three inches from the wing tip. (Bud Anderson)*

A steel ring with a clinch inside diameter was welded to the d-ring of the tow release to serve as a receptacle for a lance mounted on the wing tip of the Q-14B. The ring was mounted three inches from to: wing tip of the EC-47A. The cable for the tow release mechanism at the wing tip was connected to the standard tow release cable running to the tail of the EC-47A. The pilot of the C-47 used the standard tow release handle in the cockpit to jettison the Q-14B.

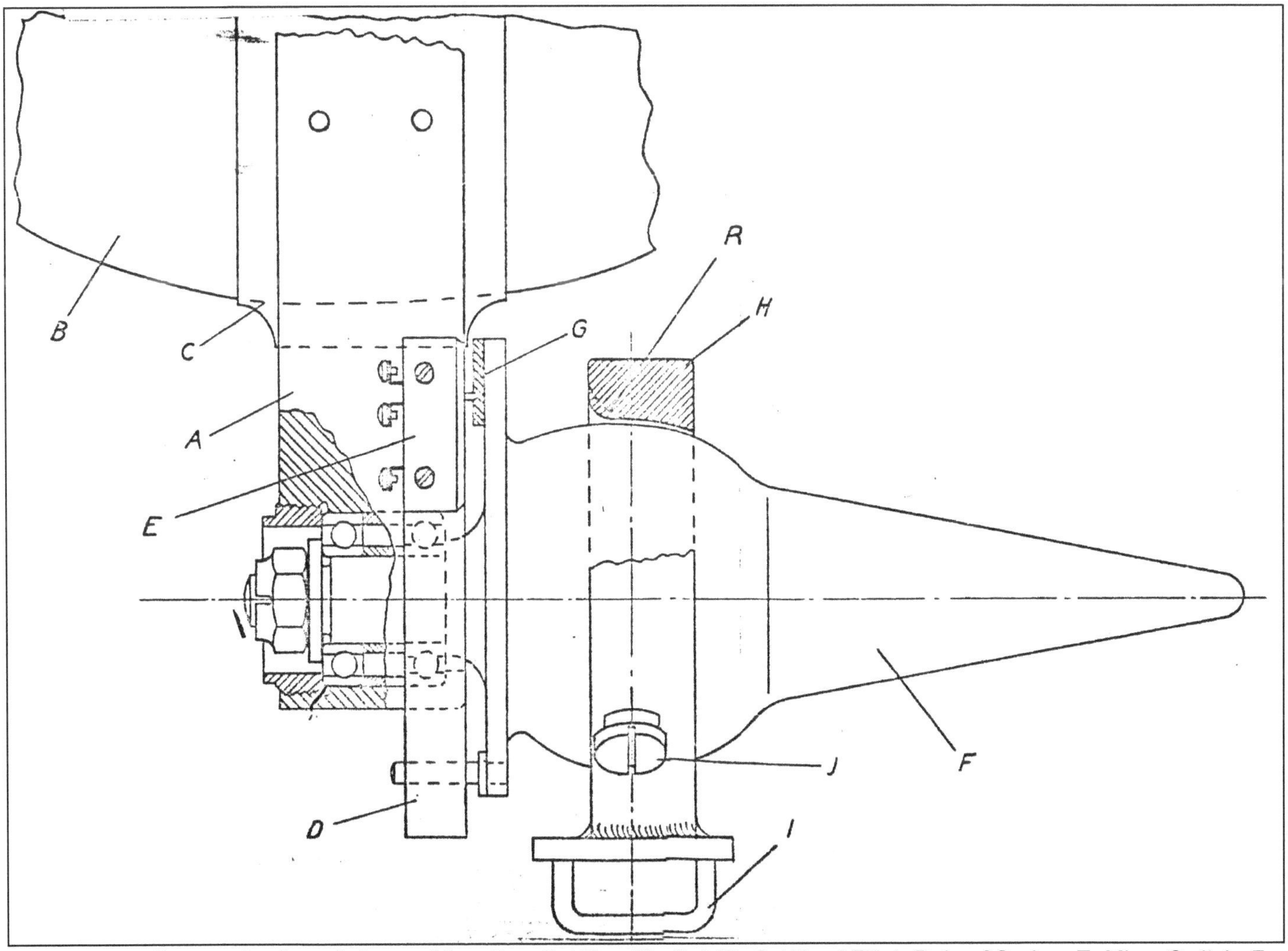


*Wing tip coupling mechanism: A. Lance Mounting, B. Wing Tip Q-14B, C. Wood Fillet, D. Leaf Spring, E. Micro Switch, F. Coupling Lance, G. Micro Switvh Actuator, H. Coupliling Ring, D-Ring, I. Engaging Doc (Bud Anderson)*

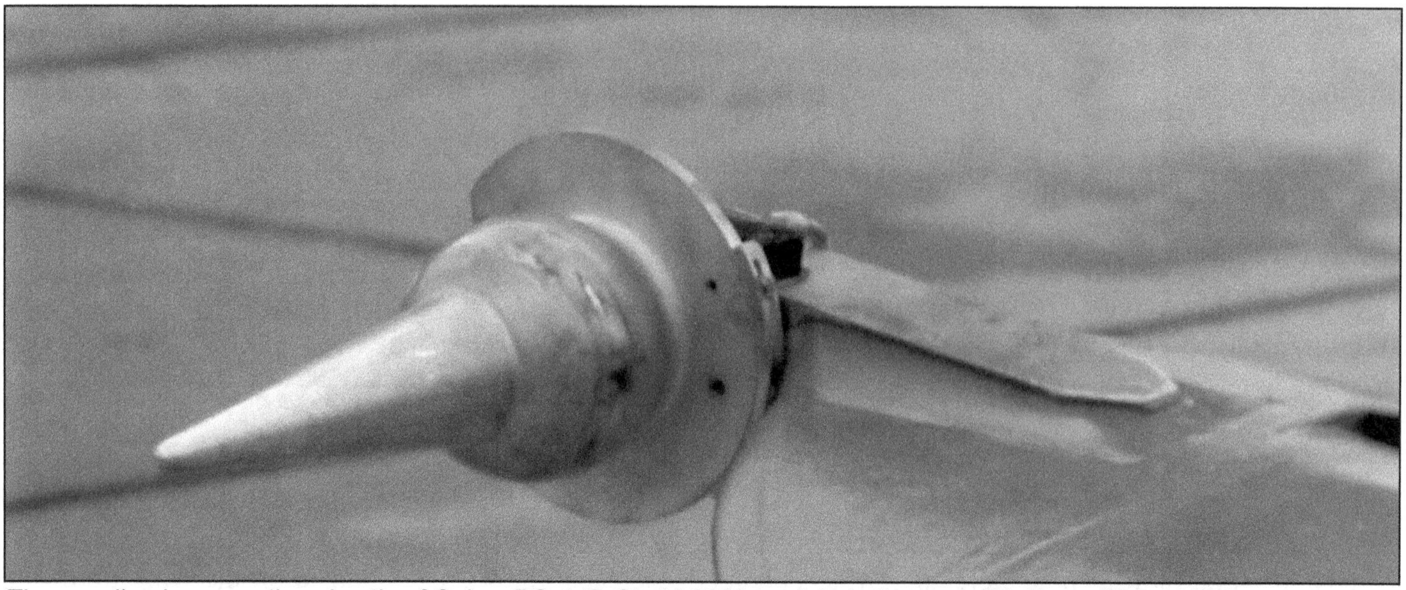

*The coupling lance on the wing tip of Culver PQ-14B-CL 44-68334 points to the rear. (Air Force Film 14425)*

*The coupling ring fit snugly wround the spherical base of the lance. It was free to rotate in pitch, yaw, and roll. (Air Force Film 14425)*

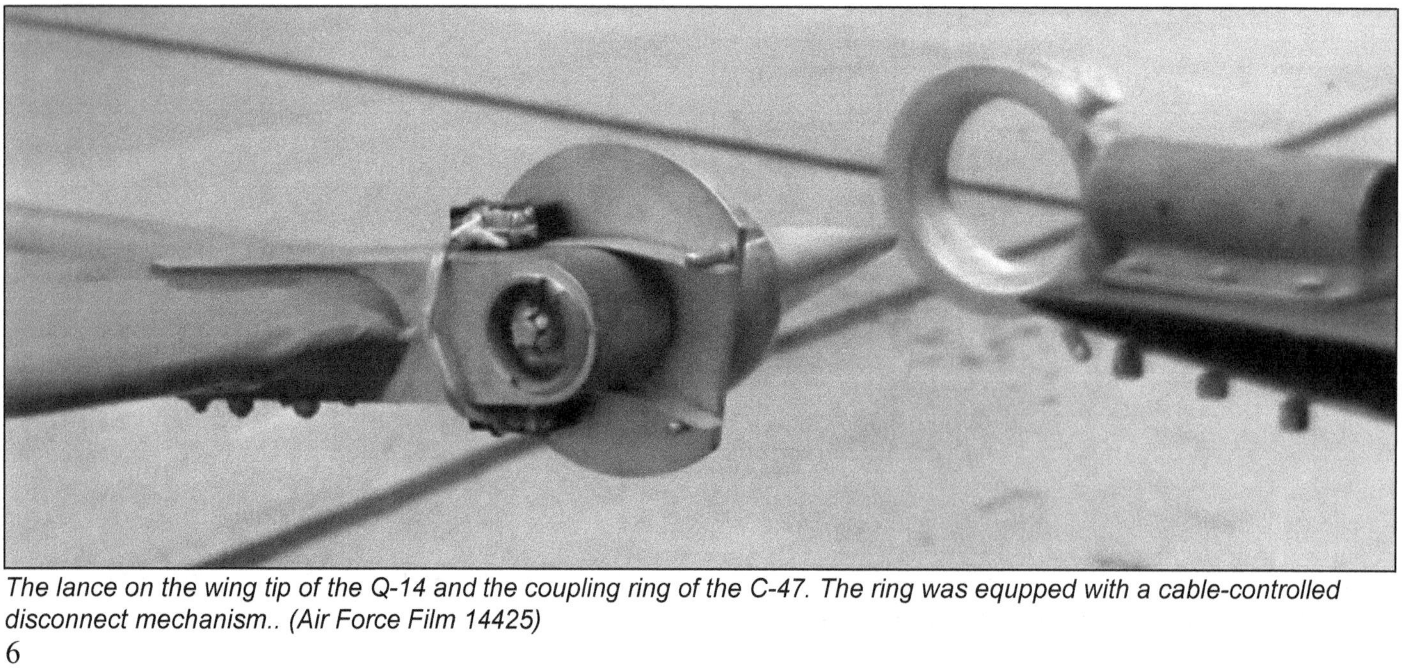

*The lance on the wing tip of the Q-14 and the coupling ring of the C-47. The ring was equpped with a cable-controlled disconnect mechanism.. (Air Force Film 14425)*

Micro switches on the back face of the lance mechanism detected the rotation of the coupling mechanism. (Air Force Film 14425)

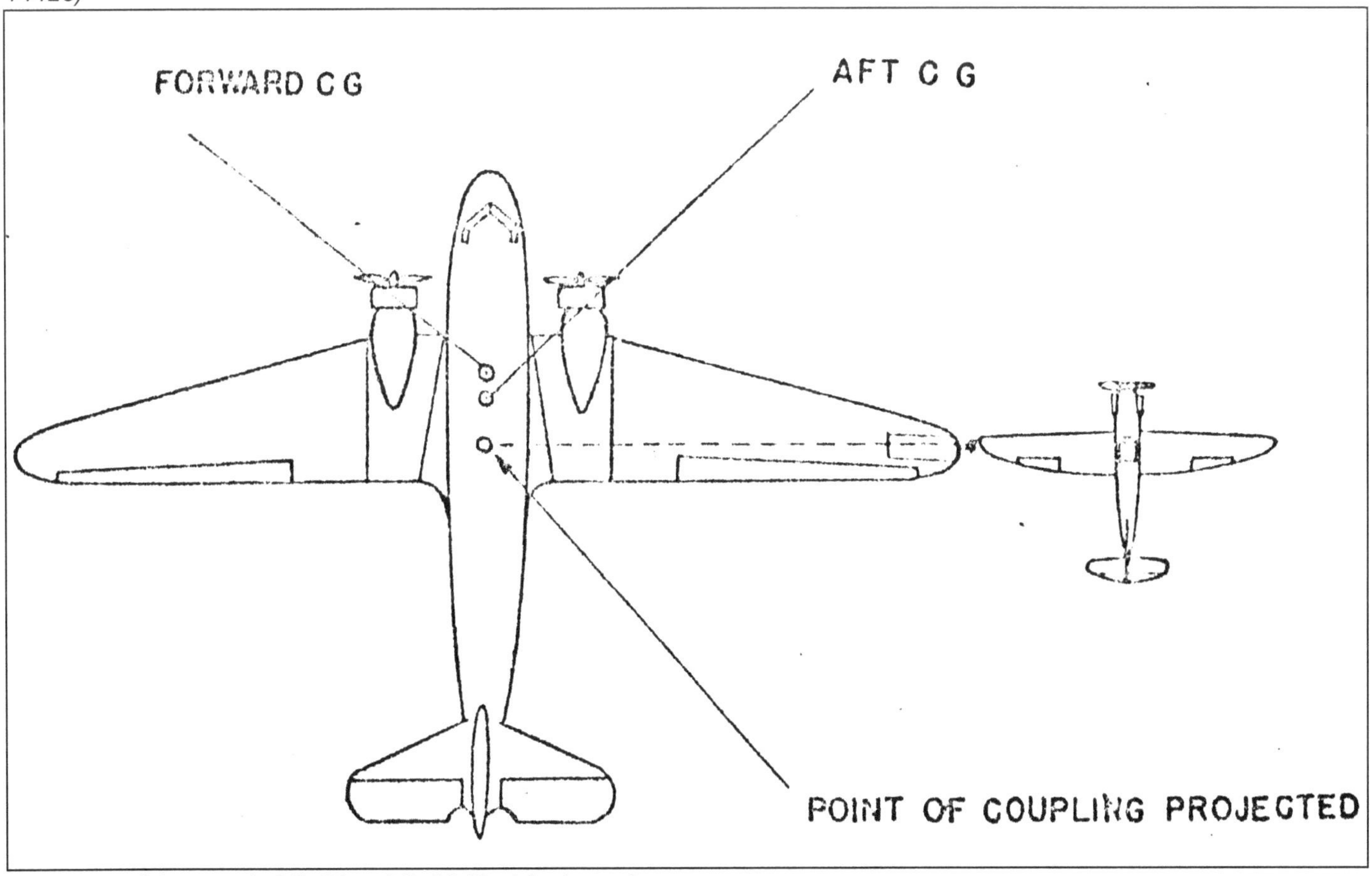


Plan view of the coupled C-47 and Q-14.(Bud Anderson)

The fifth window on the right side of the EC-47A was enlarged to provide a wide field of view for a movie camera.

The markings of the EC-47A were not significantly altered. It wore standard olive drab over gray livery from the war. There were no high-visibility colors or special mission badges applied to it.

### *Q-14B*

The Q-14B was equipped with a lance to engage the ring on the wing tip of the EC-47A. The lance pointed to the rear. The pilot of the Q-14B was expected to engage the ring by maneuvering the lance to a position in front of the ring and then backing off the throttle to seat the lance.

The lance mount was bolted to the front spar of the left wing. The Q-l4B wing was constructed as a single unit from wing tip to wing tip. Two laminated spars made of spruce or poplar ran the length of the wing. The wing ribs were constructed of-five-ply laminated poplar. They were covered with l/8-inch thick poplar plywood stressed skin. The wing leading edges, forward of the main spar, were preformed plywood. The wings incorporated fixed slots in the outboard leading edge to improve low speed control.

A hardwood fillet was glued to the wing tip on the outside of the main spar. The lance mount was forked to fit over the wing tip fillet. It was attached to the main spar with five 1/4-inch diameter flush-head aircraft-steel bolts. The bolts were countersunk to form a flush surface. A fabric covering was applied over the fillet and wing tip.

The lance was cone-shaped with a spherical base. It was constructed from solid steel. Ball bearings allowed it to rotate.

The inner surface of the ring was machined to a spherical surface with the same diameter as the base of the lance. As the base of the lance seated in the steel ring, an engaging dog in the steel ring entered a groove in the lance. The lance then rotated with the steel ring as the Q-14B rolled around the coupling. A pair of leaf springs returned the lance to the neutral position when the Q-14B was released from the EC-47A.

The lance and ring had no locking mechanism. The lance would be held in place by the drag of the Q-14B. To uncouple from the EC-47A, the O-14B pilot would advance the throttle and pull the lance out of the coupling ring. It was believed that the Q-14B pilot did not need an emergency release for the coupling mechanism.

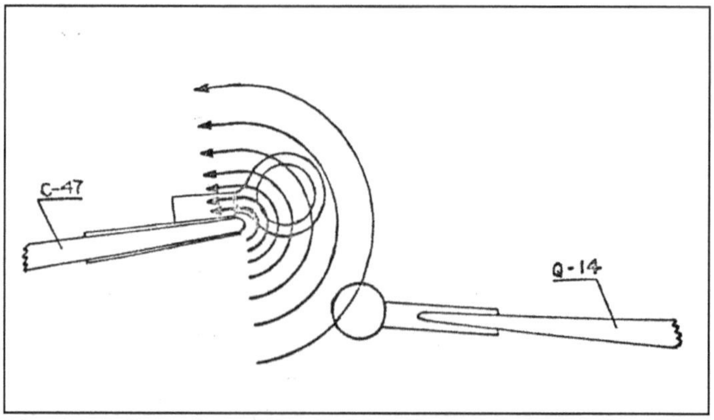

*The wing tip vortex of the C-47 deflected the wing tip of the Q-14 as it approached the ring.(Bud Anderson)*

The Q-14B was painted crimson, which was typical for drones. The bright red color warned pilots in other airplanes that there might be no pilot in the cockpit of the drone.

A large buzz-number, QK-334, was painted on the right side of the fuselage of the Q-14B. Buzz numbers made it possible to identify specific airplanes from a distance. They were composed of the last three digits of the airplane's serial number with a two-letter prefix. The first letter indicated the basic airplane type: F for fighter, B for bomber, etc. In the case of the Q-14B, the first letter of the buzz number was a Q, indicating that it was an unmanned target drone. The second letter indicated the airplane type, in this instance K for Q-l4.

### *Proposed Test Program*

The Flight Test Division of the Wright Air Development Center began its wing tip coupling investigation at Wright-Patterson Air Force Base, Ohio, in 1949. The primary objective of the tests was to determine whether the task of wing tip coupling could be accomplished at all.

The process of coupling the two airplanes was complicated by the vortex generated at the wing tip of the EC-47A. The difference in air pressure above and below any wing causes an upward flow of air around the wing tip. The rotation of the air moving around the wing tip creates a vortex that trails behind the wing. The vortex is like a small horizontal tornado.

The initial tests would determine the length of time that was required to accomplish a coupling, the best direction for the fighter to approach the coupling and the optimum diameter for the coupling ring on the EC-47A wing tip. The pilots would evaluate the stability, maneuverability, and controllability of the coupled aircraft. Observers aboard the EC-47A would

*Major Bud Anderson positions the lance of the Q-14 in front of the ring on the wing tip of the C-47 during the first test flight on August 18, 1949. (Air Force Film 14425)*

observe the attitude that the Q-14B assumed relative to the wing of the larger airplane while it was being towed. They would evaluate the ability of the pilot of the EC-47A to control the coupled aircraft in turns, climbs and descents.

The tests were not intended to quantify the improvement in range that resulted from the wing tip coupled configuration. Because the wing tip coupling mechanism was not designed to fill the gap between the wing tips of the airplanes, the full drag reduction

*Major Anderson positions the lance of the Q-14 in front of the ring on the wing tip of the C-47. (Air Force Film 14425)*

benefits of the configuration would not be realized. The relatively high air pressure below the wing forced air to flow through the gap between the wing tips, which reduced the benefit of the increased aspect ratio.

Major C. E. "Bud" Anderson was selected to be the primary pilot of the Q-14B. Anderson had downed fifteen enemy airplanes in World War II. He flew a P-51 Mustang named "Old Crow" while assigned to the

*Major Anderson prepares to seat the lance of the Q-14 in the ring on the wing tip of the C-47. (National Archives)*

357th Fighter Group, the "Yoxford Boys" of the 8th Air Force based at Leiston Field in the United Kingdom. Anderson was the leading Ace of the 363rd Fighter Squadron with 16-1/4 victories. He frequently flew as wingman to Charles E. Chuck" Yeager.

Anderson was ordered to report to the Wright Air Development Center in May 1948. His first assignment was a cold weather evaluation of the North American P-82 Twin Mustang at Ladd Field, near Fairbanks, Alaska, over the winter 1948-49. He attended the Experimental Test Pilot School at Patterson Field, Ohio and was then assigned to the Fighter Test Section.

## *Flight Test Program*

### *August 19, 1949 - First Attempted Coupling of EC-47A and Q-14*

Major Anderson practiced the method of approaching the coupling ring on August 18, 1949. Major And-arson was advised to approach the wing of the EC-47A frown below. The upward flow around the wing tip of the EC-47A would then serve to direct the lance toward the coupling ring.

During his first approach to the wing tip of the EC-47A, Major Anderson studied the effect that its wingtip vortex had on the Q-14B. The upward flow of air imparted a strong outward rolling force on the Q-14B. Approaches were flown at various airspeeds and altitudes to determine the optimum conditions. Anderson's initial impression was that twit would be quite easy to make the hook-up."

On the morning of August 19, Major Anderson took off from Wright Field in the Q-14B to make the

*Major Anderson seats the lance of the Q-14 in the ring on the wing tip of the C-47. (Air Force Film 14425)*

*The Q-14 immeditely rolled to the right and fell away from the C-47. (Air Force Film 14425)*

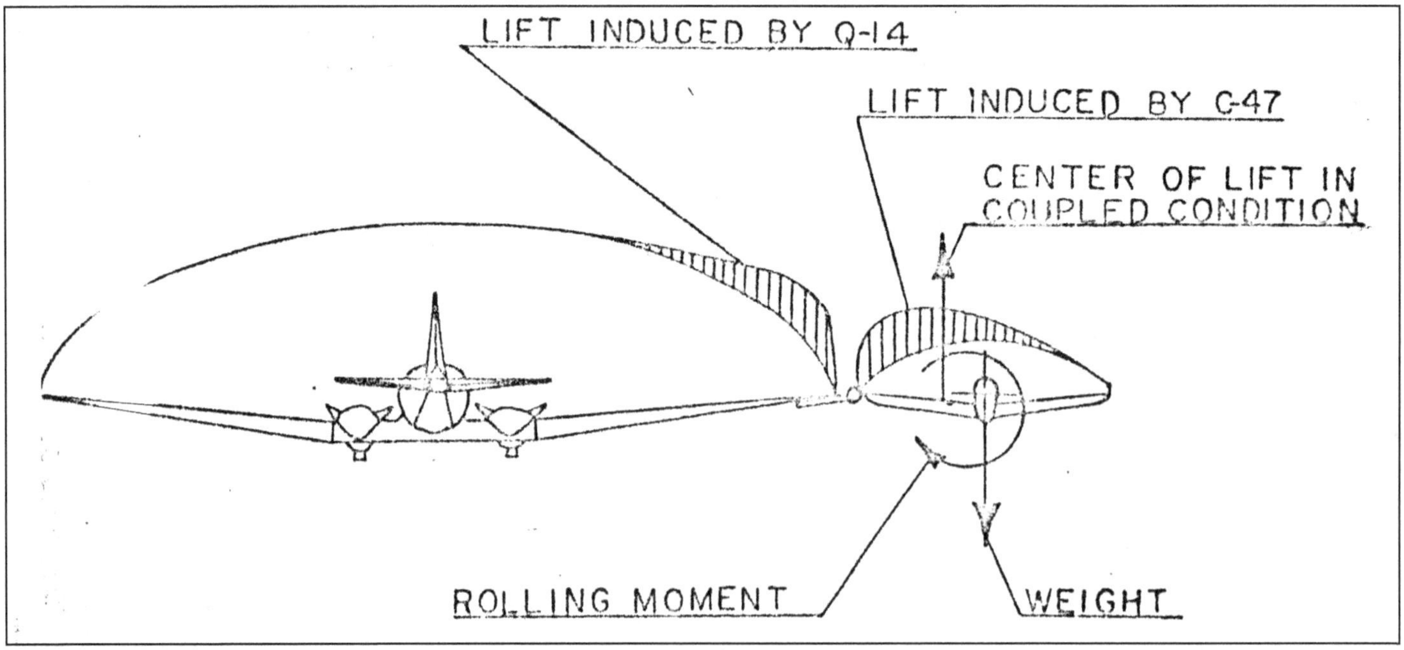


*Shift in the lift distribution of the coupled C-47 and Q-14. (Bud Anderson)*

*Major Bud Anderson positions the lance of the Q-14 in front of the ring on the wing tip of the C-47 during the first test flight on August 18, 1949. (Air Force Film 14425)*

*Major Anderson positions the lance of the Q-14 in front of the ring on the wing tip of the C-47. (Air Force Film 14425)*

observe the attitude that the Q-14B assumed relative to the wing of the larger airplane while it was being towed. They would evaluate the ability of the pilot of the EC-47A to control the coupled aircraft in turns, climbs and descents.

The tests were not intended to quantify the improvement in range that resulted from the wing tip coupled configuration. Because the wing tip coupling mechanism was not designed to fill the gap between the wing tips of the airplanes, the full drag reduction benefits of the configuration would not be realized. The relatively high air pressure below the wing forced air to flow through the gap between the wing tips, which reduced the benefit of the increased aspect ratio.

Major C. E. "Bud" Anderson was selected to be the primary pilot of the Q-14B. Anderson had downed fifteen enemy airplanes in World War II. He flew a P-51 Mustang named "Old Crow" while assigned to the

*Major Anderson prepares to seat the lance of the Q-14 in the ring on the wing tip of the C-47. (National Archives)*

357th Fighter Group, the "Yoxford Boys" of the 8th Air Force based at Leiston Field in the United Kingdom. Anderson was the leading Ace of the 363rd Fighter Squadron with 16-1/4 victories. He frequently flew as wingman to Charles E. Chuck" Yeager.

Anderson was ordered to report to the Wright Air Development Center in May 1948. His first assignment was a cold weather evaluation of the North American P-82 Twin Mustang at Ladd Field, near Fairbanks, Alaska, over the winter 1948-49. He attended the Experimental Test Pilot School at Patterson Field, Ohio and was then assigned to the Fighter Test Section.

## Flight Test Program

### August 19, 1949 - First Attempted Coupling of EC-47A and Q-14

Major Anderson practiced the method of approaching the coupling ring on August 18, 1949. Major And-arson was advised to approach the wing of the EC-47A frown below. The upward flow around the wing tip of the EC-47A would then serve to direct the lance toward the coupling ring.

During his first approach to the wing tip of the EC-47A, Major Anderson studied the effect that its wingtip vortex had on the Q-14B. The upward flow of air imparted a strong outward rolling force on the Q-14B. Approaches were flown at various airspeeds and altitudes to determine the optimum conditions. Anderson's initial impression was that twit would be quite easy to make the hook-up."

On the morning of August 19, Major Anderson took off from Wright Field in the Q-14B to make the

*Major Anderson seats the lance of the Q-14 in the ring on the wing tip of the C-47. (Air Force Film 14425)*

*The Q-14 immeditely rolled to the right and fell away from the C-47. (Air Force Film 14425)*

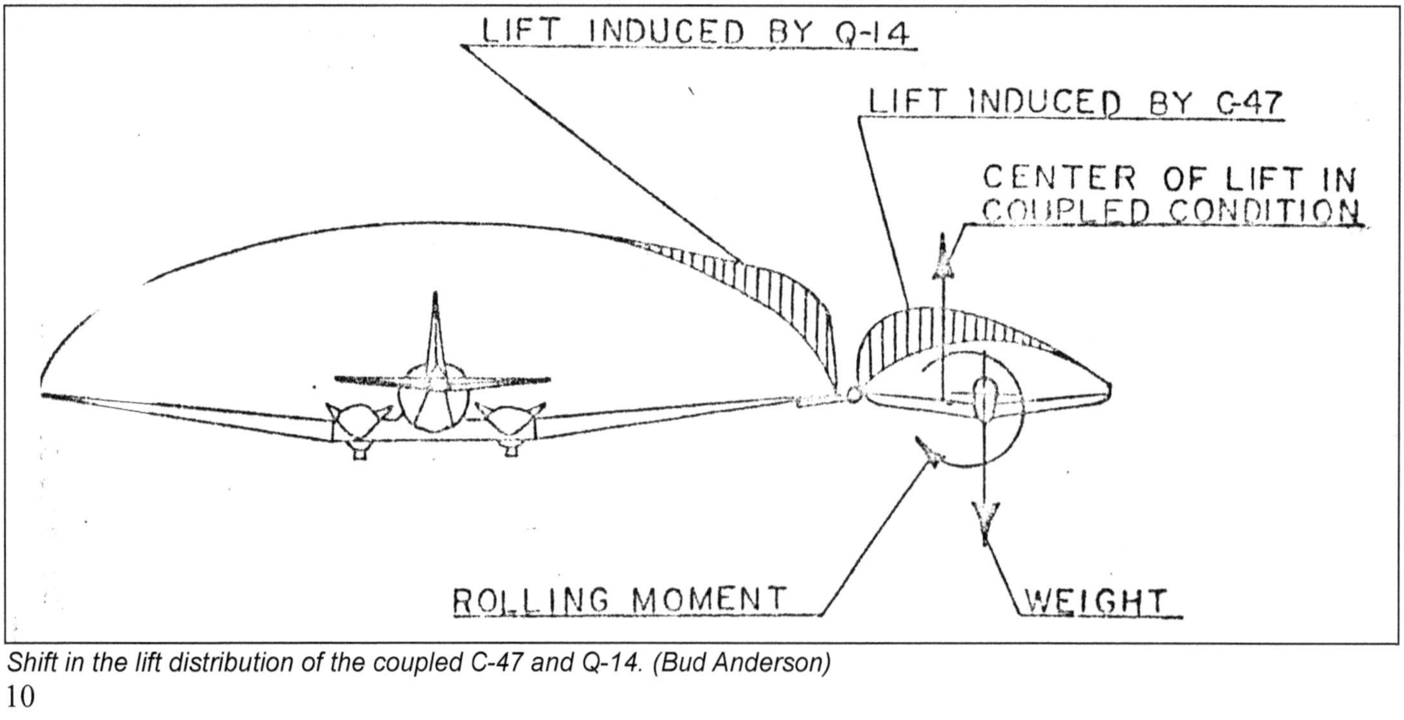


*Shift in the lift distribution of the coupled C-47 and Q-14. (Bud Anderson)*

first attempt to couple it to the wing tip of the EC-47A. Walt Rosenfield piloted the EC-47A. He stabilized the EC-47A in straight and level flight at 8,000 feet altitude. Anderson flew the Q-14B to a position ahead of the right wing tip of the EC-47A. At that altitude, the Q-14B was flying at nearly full throttle to keep up with the EC-47A. The drone's engine had little power in reserve for maneuvering.

Anderson throttled back to thread the lance into the ring, but then he discovered that the vortex streaming from the wing tip of the EC-47A made it difficult to make a steady approach to the ring. His wing tip was forced up and outward, away from the coupling ring, as he moved back relative to the EC-47A. Anderson spent half an hour making repeated attempts to link the airplanes together. The difficulty in making contact with the coupling ring raised concern in his mind about the controllability of the Q-14B if he should succeed in coupling the planes together.

After several attempts, Anderson positioned the lance directly in front of the ring and chopped his throttle. The lance engaged the ring. Immediately, the Q-14B dropped off to the right. It rolled nearly 90 degrees to a wings-vertical attitude. Although Anderson moved the control stick all the way to the left, the ailerons did not provide enough control authority to prevent the Q-14B from dropping.

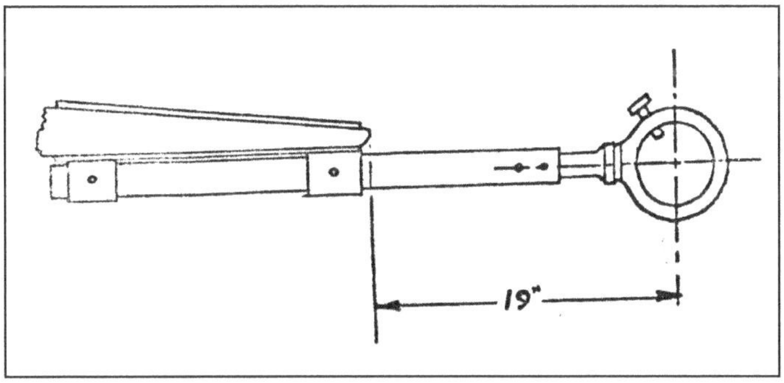


*Modified coupling ring on the C-47 wing tip. The center of teh ring was positioned nineteen inches from the wing tip.(Bud Anderson)*

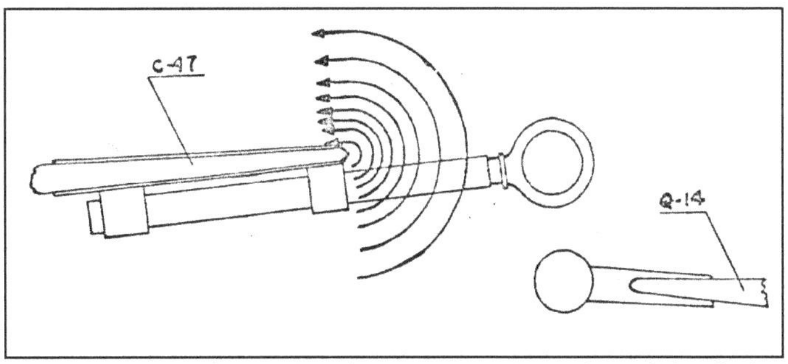


*The influence of the wing tip vortex on the wing of the Q-14 was reduced by moving the ring farther from the wing tip. (Bud Anderson)*

*The coupling ring was extended farther from the C-47 wing tip. (Air Force Film 14425)*

*The Q-14 and the modified coupling ring of the C-47. (Air Force Film 14425)*

Anderson applied full throttle to extract the lance from the ring and pulled the Q-14B away from the EC-47A. Less than five seconds elapsed from contact to release. The wing tip of the EC-47A was slightly bent during the incident, so no more attempts were made to link up on that flight.

Analysis of the lift distribution of the coupled airplanes revealed that the center of lift over the Q-14B moved toward the EC-47A as the wing tips came together. With the center of lift of the Q-14B offset from the center of mass, it tended to roll downward around the wing tip coupling. It had not been realized that the ailerons of the Q-l4B would provide insufficient control authority to overcome this tendency.

The center of lift over the wing of the EC-47A also moved toward the Q-14B, but the displacement was much smaller. The pilot of the EC-47A was able to easily adjust the trim of the airplane to compensate for the effect.

In preparation for the second mating attempt, additional reinforcement was installed on the wing tip of the EC-47A. l/4-inch thick dural plates, measuring 15 inches by 63 inches, were bolted to the top and bottom of the wing.

The coupling ring was mounted on an extended boom that positioned the ring out of the strongest part of the wing tip vortex. It was hoped that positioning the ring in the smoother air farther from the wing tip would make it easier for Anderson to make the link-up.

Two steel boom brackets were mounted on the underside of the wing. A 2 3/4-inch diameter steel pipe was mounted in the brackets. A glider tow release was attached to the end of the pipe. The-fixture allowed the distance between the wing tip and the receptacle to be adjusted. When the boom was fully extended, the center of the steel ring was 19 inches from the wing tip. Four holes were drilled through the boom to enable technicians to conveniently fix the ring at specific distances from the wing tip.

#### *October 7, 1949 - second flight*

The second wing tip coupling flight was made on October 7, nearly two months after the first attempt. Anderson still had difficulty lining the lance up with the coupling ring even with the ring in the extended position. It took considerable maneuvering to engage the lance in the ring.

Anderson did not back off on the throttle when the lance engaged the ring as much as he had on the previous flight. He found that he was still not able to counteract the rolling tendency of the Q-14B using the ailerons alone.

A little experimentation revealed that the elevator was far more effective at maintaining a wings-level attitude than the ailerons were. Anderson was able to keep the Q-14B level by pulling back on the stick when the little plane dropped off to the right. It required some mental effort to switch his roll control inputs to the elevator, which went against his conditioned reflexes.

*Major Anderson made the first successful coupling on October 7, 1949.. (Air Force Film 14425)*

Hooking the Q-14B to the wing tip of the EC-47A moved its roll axis away from its usual position between the ailerons to a position set off to one side of both ailerons. This caused the inboard aileron to apply its control force in the opposite direction of the desired rotation, making the ailerons much less effective as roll control devices.

The solution to the roll control problem was to adjust the amount of lift generated by the wing by applying the elevator to change its angle of attack. While this provided effective control of the flapping angle[2] of the Q-14B, it also changed the pitch angle of the Q-14B relative to the wing of the EC-47A.

The spherical mating surfaces of the ring and lance coupling allowed the Q-14B to rotate in pitch without transmitting any twisting force to the wing of the EC-47A.

Anderson spent a few minutes getting a feel for the behavior of the Q-14B while coupled to the wing tip of the EC-47A.

He then advanced the throttle and disengaged from the coupling ring. He lined up for another attempt and successfully re-engaged the ring.

Anderson experimented with adjusting the throttle of the Q-14B as it was towed by the EC-47A. He found that the elevator provided enough control authority for him to throttle the engine back to idle. The Q-14B yawed away from the EC-47A as the power was reduced, but quickly stabilized at a small yaw angle. The little airplane could be pointed straight ahead by the application of left rudder. Most of the time, the rudder was left in the neutral position, and the Q-14B was allowed to yaw outward while coupled to the wing of the EC-47A.

Anderson threaded the lance into the ring four

*Major Anderson made the first successful coupling on October 7, 1949. (Air Force Film 14425)*

2 The angle of displacement of the Q-14B from a position in line with the wing of the EC-47A is referred to as the "flapping angle."

*The Q-14 coupled to the wing of the C-47 in flight. Panchromatic film makes the crimson Q-14 appear light in pictures taken by the chase plane.(Bus Anderson)*

times. The longest coupled flight lasted five minutes. The total time that the airplanes were coupled together on this flight was ten minutes.

In short order Anderson acquired confidence in his ability to make the link-up and maintain the proper attitude in coupled flight. He found that it was easy to accomplish the coupling in less than half a minute. Once the lance was lined up with the coupling ring, a small decrease in throttle setting was adequate to decelerate the Q-14B and engage the coupling mechanism. [iii]

#### *Flight characteristics Evaluation, October 10, 11*

The third wing tip coupling flight was flown on October 10. Anderson concentrated on evaluating the control of the Q-14B with its engine idling as it was towed by the EC-47A. He made only three contacts, but he accumulated forty minutes of towed flight time. The next day he coupled the Q-14B to the EC-47A for an hour during two periods of coupled flight.

Walt Rosenfield evaluated the flight characteristics of the EC-47A with the Q-14B coupled to its wingtip. He increased the speed of the coupled airplanes to 126 males per hour. The stability of the Q-14B improved as the speed increased. The handling characteristics of the coupled airplanes also improved at higher speed. Rosenfield found that steep turns with bank angles up to 50 degrees could be flown quite easily.

Anderson found that it was not possible to fly the Q-14B hands-off while it was coupled to the wing of the EC-47A. Any yawing movement of the EC-47A caused the Q-14B to roll up and down about the coupling mechanism. The oscillation increased until the Q-14B pilot stopped it.

*The elevator was the primary control for maintaining the position of the Q-14 realtive to the wing tip of the C-47.(Air Force Film 14425)*

#### Lower Altitude, October 20

It was quickly concluded that future couplings should be conducted at the lowest altitude where the air was smooth enough. The EC-47A needed to fly at an indicated airspeed of at least 95 knots to maintain stability. The Q-14B had to use nearly full throttle to go that fast at 8,000 feet altitude.

For the next wing tip coupling flight, flown on October 20, the coupling altitude was moved down to 4,500 feet. At lower altitude the Q-l4B could maintain formation with the EC-47A at a lower throttle setting. This left a greater reserve of power for maneuvering relative to the larger airplane.

Testing determined that the best airspeed for coupling was between 110 and 115 miles per hour. The optimum altitude for coupling was between 4,000 and 6,000 feet, due to the limited power of the Q-14B.

#### Cruising Coupled, November 1949

When Anderson had established the feasibility of coupling the Q-14B to the wing tip of the EC-47A in flight, the emphasis of the tests shifted to evaluating cruising flight with the airplanes coupled together. Anderson flew eight wing tip coupling flights in November. One to three couplings were accomplished on each flight. Coupled flight time on each flight ranged from 21 minutes to 1 hour and 23 minutes.

*The Q-14 entered a diverging dutch roll in hands-off flight with the wing tips coupled.(Air Force Film 14425)*

Flight through clouds while the airplanes were coupled was first demonstrated on the seventh flight of the project on November 2. It was felt that coupled flight through clouds was less demanding than close formation flight through clouds.

The EC-47A could accelerate to higher speed once the Q-14B was attached to its wing tip. The coupled aircraft became more stable at 120 knots, which was closer to the normal cruising speed for the EC-47A.

It was necessary to adjust the trim of the EC-47A to fly straight and level at 130 miles per hour with the Q-14B engine idling. The test team established that the EC-47A required six degrees of right wing down aileron trim to compensate for the lift generated by the Q-14B. It was necessary to crank in 2.5 degrees left rudder trim to compensate for the drag of the small airplane. Asymmetric trim would be less of a concern to the full-scale project because most of the time there would be an F-84 *Thunderjet* mounted on each wing tip of a B-29 Superfortress.

The turn-and-bank indicator became the primary attitude instrument for its pilot. Anderson's elevator control inputs were guided by the watching the ball in the indicator. When it moved right, he would pull back on the stick. When it moved left, he would push forward on the stick. There was no need to look out of the cockpit to watch his position relative to the EC-47A. It made an odd impression on the crew of the EC-47A to watch him maintain his position on the wing tip without bothering to look at the bigger plane.

#### Adjust Elevator Trim, 16 November 1949

The elevator trim tab of the Q-14B was doubled in size to increase its effectiveness for the next flight, flown on November 16. The horizontal stabilizer setting was adjusted downward by 2 degrees to increase the elevator nose-up trim of the airplanes Following these adjustments to the elevator trim, Anderson could trim the Q-14B to fly straight and level on the wing tip of the EC-47A with minimal control force. The reduction in pilot control forces meant that long endurance cruise flights could be flown.

#### Reversed Coupling, 18 November 1949

The wing tip coupling mechanism intended for

*Coupled flying time reached a total of thirty hours. (Bus Anderson)*

use in the full-scale B-29/F-84 flight tests was different than the system that equipped the Q-14B and EC-47A. The lance designed for use on the F-84 faced forward, so a series of flights was flown with the lance on the Q-14B mounted facing forward to evaluate the task of approaching the coupling ring from the rear.

The first flight with the lance facing forward was flown on November 18. Earlier that day, Major Anderson made a flight with the lance in its normal position. He coupled the Q-14B to the EC-47A three times and flew for 45 minutes while attached to its wing tip.

After the first flight of the day, the lance was reversed to face forward. The two airplanes took off again and Anderson demonstrated that it was possible to maneuver the lance into the coupling ring from the rear. He made a single contact and remained coupled to the EC-47A for three minutes. Because there was no locking mechanism, Anderson had to hold full throttle to keep the lance in the receiver. The coupling mechanism for the B-29/F-84 tests would incorporate a latch to hold the F-84 in place when its engine was throttled down. The test demonstrated that coupling the planes with a forward facing lance was feasible and paved the way for the full-scale wing tip coupling project.

***More pilots, December 1949***

As confidence grew in the feasibility of wing tip coupling, additional pilots were assigned to evaluate

*The combination was flown at airspeeds between 100 and 160 miles per hour. (Air Force Film 14425)*

the concept. The team selected pilots with experience in a variety of different airplane types to address the question of what constituted suitable preparation for the task of wing tip coupling. The pilots' training was evaluated based on the amount of flight time that each pilot had accumulated, broken into the categories of single-engine, twin-engine, multi-engine, and jet.

Captain Darrell made the first wing tip coupling flight by a pilot other than Major Anderson on December 1. He had over 1000 hours of flight time in twin-engine airplanes and 700 hours in single-engine airplanes. He had very little multi engine experience. He had flown 389 hours in jets. He made wing tip coupling flights on December 1, 2 and 5. On his third flight on December 5, the lance was mounted facing forward.

Major Anderson flew another flight to evaluate the forward facing lance on December 7.

#### *First Night Flight, 13 December 1949*

On December 13, Walt Rosenfield evaluated the behavior of the coupled pair at very low altitude. He descended to an altitude just 250 feet above the ground. This was the lowest altitude at which the coupled aircraft were flown during the tests. Anderson made nineteen contacts with the coupling ring during 21 minutes of coupled flight.

That night Anderson made the first night wing tip coupling flight. A special lighting system was installed in the airplanes to assist the Q-14B pilot during night couplings. The lights were aimed to illuminate only the lance and the receiving ring to reduce unnecessary reflections from other parts of the airplanes. A 50-watt, sealed-beam spotlight was installed on the coupling boom of the EC-47A approximately 12 inches behind the coupling ring. Its light beam was directed through the ring to illuminate the lance when it was in the proper position. The switch that operated the spotlight was located on the co-pilot's instrument panel.

A 21-candle-power light was installed pointing rearward on the left wing tip of the Q-14B. A rheostat was installed in the cockpit of the Q-14B to allow its pilot to control the brightness of the wing tip light.

Major Anderson made 26 contacts with the coupling ring during the first night wing tip coupling flight on December 13. Approaches were made with various lighting arrangements. It was found to be necessary for both wing tips to be illuminated to successfully engage the lance in the coupling ring. Turning on the cargo compartment lights in the EC-47A provided Anderson with a better perspective of his position relative to the larger airplane. Anderson spent 69 minutes flying coupled to the wing tip of the EC-47A that night.

Major Anderson did not consider night link-ups to be especially difficult, but the pilots of both airplanes felt that the light from the spotlights impaired their night vision.

Another flight was flown the following night. In preparation for the flight, visors were installed on the spotlights to restrict their field of illumination to the coupling mechanism on the wing of the other airplane. Dark matte-finish paint was applied to the bright metal surfaces of the coupling to reduce glare from the lights.

#### *Additional Pilots, 14 December 1949*

Two new pilots were introduced to the technique of wing tip coupling on December 14. Major Frank K. Everest, the assistant chief of the fighter test section, made three contacts with the coupling ring and flew the Q-14B coupled to the wing tip of the EC-47A for 25 minutes. From March to September 1949, Major Everest had been flying the Bell X-1 at Edwards AFB. In April 1950 he would transfer back to Edwards Air Force Base.

Major Charles E. Yeager also made three contacts with the coupling ring on his familiarization flight that day. Major Yeager was assigned to Edwards Air Force Base, and his flying records were not available to the test team. The project report noted, "it is known that he has extensive fighter flying experience."

#### *Second Night Flight, 14 December 1949*

Anderson made another night wing tip coupling flight the evening of December 14. In the course of the two-hour night flight, he contacted the coupling ring thirty times.

The team concluded that it would be advisable

*The Q-14 coupled to the wing of the C-47 over Wright Field. (Bus Anderson)*

to install several reference lights on the fuselage and leading edge of the wing of the larger airplane to be used for any future night flights to improve the perspective of the pilot of the attaching airplane. They also recommended that a rheostat should be installed to control the brightness of the spotlight on the wing tip of the EC-47A.

#### *Additional Pilots, 15 December 1949*

On December 15, Lt. Colonel Richard L. Johnson, the Chief of Fighter Test at Wright Field, made his first wing tip coupling flight in the Q-14B. Johnson's experience was primarily in single-engine airplanes. He had 767 flight hours in jets, the most of any of the pilots to fly the Q-14B during 1949.

Another new pilot, Lt. Colonel R. F. Schleech also made a wing tip coupling flight that day. Schleech's flying experience was mostly in multi-engine airplanes. He had 278 hours of jet flight time.

In the space of three days, five pilots made seven wing tip coupling flights. A total of 84 couplings were accomplished. The Q-14B was towed by the EC-47A for nearly four hours. Following this flurry of activity, the wing tip coupling project went into hiatus for two months.

#### *Minimum Ring Distance, February 3, 1950*

When testing resumed on February 3, 1950, Anderson's confidence in his ability to make the link-up prompted the test team to move the coupling ring back to its original position, right next to the wing tip of the EC-47A. Anderson determined that an experienced and confident pilot could make the link-up at the closer position.

#### *Q-14 Attitude*

The team investigated the change in the flight attitude of the Q-14A as it approached the wing tip of the EC-47A. A 1-inch wide white line was painted along the left side of the Q-14B to provide a reference for measuring the pitch angle of the airplane as it flew alongside the EC-47A. An adjustable platform with a protractor for angle measurement was mounted inside the fuselage of the EC-47A.

The measurements were made while the airplanes were flying at an air speed of 120 miles per hour and an altitude of 7,000 feet. Measurements were taken

*Turns were made in both directions at angles of bank up to 50 degrees. (Air Force Film 14425)*

*Multi-engine pilots were less likely to successfully engage the coupling mechanism.(Air Force Film 14425)*

with the Q-14B flying at three different distances from the wing tip of the EC-47A. The relative pitch angle of the Q-14B was measured when it was about 30 feet away from the wing tip as a baseline value. Anderson approached to within 2 feet of the wing tip for the next measurement. He then engaged the coupling ring for the third measurement.

When the Q-l4B was flying two feet from the wing tip of the EC-47A, the extra lift induced by the proximity to the larger airplane allowed it to fly at a lower angle of pitch. Its nose lowered by about one degree. Once the two airplanes were coupled, the application of elevator to maintain the proper flapping angle caused the nose of the Q-14B to pitch up to an angle slightly higher than it had assumed when it was flying 30 feet away.

### Center of Gravity Limits

The load in the EC-47A was adjusted to move the center of gravity to the aft limit for the next test flight on February 7. Walt Rosenfield investigated the behavior of the coupled airplanes while the EC-47A was tail heavy. Rosenfield banked the airplane to each side at angles up to 50 degrees. He determined that the Q-14B had a negligible effect on the motion of the EC-47A even when Major Anderson rolled it up and down around the wing tip.

Lt. Colonel Johnson flew the Q-14B again on February 20. For this test, the center of gravity of the EC-47A was adjusted to its forward limit. Rosenfield evaluated the behavior of the coupled airplanes while the EC-47A was nose-heavy. The EC47A was less affected by the presence of the Q-14B with the center of gravity in this position. Moving the center of gravity forward increases the distance between it and the stabilizing surfaces at the rear of the airplane, increasing the stability of the airplane slightly.

In the five days from February 20 to February 24, four different pilots flew wing tip coupling flights in the Q-14B. Eleven couplings were accomplished. The EC-47A towed the Q-14B for 2 hours and 15 minutes.

### Multi-engine Pilots

Lt. Colonel W. P. Maiersperger, the wing tip coupling project manager, made his first wing tip coupling on February 21. Most of his 2,000 hours of flying experience was in twin-engine airplanes, but he had nearly 800 hours in single-engine airplanes. He had only six hours of experience flying jets.

Maiersperger coupled the Q-14B to the EC-47A once that day and was towed for 25 minutes. On February 23, he made 6 contacts with the coupling ring and was towed for 44 minutes.

Captain T. M. Christenson was introduced to the task of wing tip coupling on February 23. He was a transport pilot with over 2300 hours in multi-engine airplanes. He had only 32 hours in single-engine airplanes and no time in fighters or jets.

In the course of the two-hour flight, Christenson was not able to engage the coupling ring with the lance. He had difficulty flying formation off the wing tip of the EC-47A. He found that the elevator of the Q-14B was too sensitive. As he lined up the lance with the coupling ring, he over-controlled the airplane in pitch, oscillating up and down. His reflexes were conditioned to the slower response of a larger airplane.

On February 24, Captain A. Vincenzi made his first wing tip coupling flight. He had over 3,000 hours of flight time in twin- and multi-engine airplanes, but just 134 hours in single-engine airplanes and 34 hours in fighters.

*It was easier to fly with the airplanes coupled, than to fly in close formation uncoupled. (Bus Anderson)*

Vincenzi was challenged by the task of maintaining position off the wing tip of the EC-47A. The observers in the EC-47A informed him over the radio that he was over-controlling the elevator of the Q-14B. After one hour and ten minutes of bobbing about off the wing of the EC-47A, he snagged the coupling ring with the lance.

Shortly after Vincenzi engaged the ring, he over-controlled the elevator, pitching the Q-14B up until it rolled to a nearly wings vertical attitude. The co-pilot of the EC-47A yanked on the tow release handle, but the deflection of the wing caused the cable to bind, preventing the release of the Q-14B. Captain Vincenzi attempted to return to a wings level attitude, but he overshot and the Q-14B dropped below the wing of the EC-47A. The little plane dragged the right wing of the EC47A down, and the coupled aircraft entered a steep right turn. The wing tip of the EC-47A buckled downward and over six feet of the right aileron separated from the airplane.

The co-pilot pulled on the glider tow release handle twice more to free the coupling ring from the boom. The ring separated from the EC-47A, freeing the Q-14B, and then fell from the lance. Both airplanes landed safely back at Wright-Patterson Air Force Base.

A month passed before the next wing tip coupling flight was flown. Following the incident on February 24, a new receiver ring of larger diameter was fabricated. It had been discovered that relatively small side loads could prevent the release of the lance from the original small diameter ring. Initially, the tow release cable ran through steel tubing inside the wing of the EC-47A, but high friction forces had interfered with its operation. The tow release cable was replaced with Bowden Cable to reduce the friction and ensure proper operation of the release mechanism. On the basis of the difficulties encountered releasing the Q-14B, the design of the emergency jettison system to be used in the full-scale project was revised.

In all, nine pilots flew wingtip-coupling flights during the fall and winter of 1949-50. Two of the nine pilots were not able to accomplish the link-up. The

*From this series of tests, it was concluded that it is possible to couple aircraft in flight at their wing tips.(Air Force Film 14425)*

pilots with primarily multi-engine experience and limited light plane experience were the least able to master the technique.

A chase-plane accompanied the EC-47A and Q-l4B on March 30 to make movies of the wing tip coupling operation. Major Anderson made seven contacts with the coupling ring, and the EC-47A towed the Q-14B for one hour and eighteen minutes. At one point during the filming Anderson raised both hands in the air to show that the airplane could be trimmed to fly hands-off for a short while. In fact, the Q-14B would begin a dutch-roll of increasing amplitude if the pilot left the controls free for any length of time.

#### *Single Aileron, April 1950*

During coupled flight, the ailerons were ineffective for correcting the rolling of the Q-14B around the wing tip of the EC-47A. When the pilot moved the stick to the left to raise the Q-14B, the left aileron moved upward, applying downward force that opposed the desired movement.

In early April 1950, the left aileron of the Q-14B was disconnected to see whether the right aileron operating alone would keep it level with the wing of the EC-47A. Several flights were made to establish the proper position in which to fix the left aileron of the Q-14B. Without the neutralizing force of the linkage to the left aileron, the right aileron floated up nine degrees from the neutral position, reducing the lift from that wing. The 22-pound lance installation on the left wing tip compensated for part of the asymmetric lift. Fixing the left aileron three degrees up from the neutral position eliminated the rest of the control asymmetry. Anderson could trim the Q-14B to fly hands-off with its wings level while flying separate from the EC-47A.

Aileron effectiveness was evaluated by observing the greatest rate of roll of the EC-47A that the Ql4B could follow. Due to the lack of an available rate of roll indicator, the team used a sensitive rate of pitch indicator, turned 90 degrees, to measure the roll rate of the EC-47A to a precision of two degrees per second.

When Anderson used the elevator as the primary control, the Q-14B could follow the EC-47A at roll rates up to eight degrees per second. When using the ailerons alone, rigged in the normal fashion, it was not possible for the Q-14B to follow any maneuver performed by the EC-47A.

Anderson coupled the Q-14B to the wing tip of the EC-47A to evaluate the effectiveness of the single-aileron arrangement on April 7. Flight control effectiveness using just the outboard aileron was marginally better than when they were rigged normally. The right aileron acting alone was just barely able to control the Q-14B. The Q-14B was unable to follow rates of roll greater than 2 degrees per second. Anderson concluded that the control authority of the right aileron could easily be overwhelmed by a small displacement of the elevator.

#### *Long Endurance Flight, 15 April 1950*

Anderson piloted the Q-14B coupled to the wing tip of the EC-47A for 4 hours and 8 minutes on a round-robin flight on April 15. The inboard aileron remained fixed in place for this flight. In the course of that flight they flew over Cleveland, Toledo, and Indianapolis, Indiana. Due to the cold weather, Anderson spent much of the flight hunched down in the cockpit with his parka pulled over his head, just watching the ball of the turn-and-bank indicator.

At the conclusion of the endurance flight, Anderson made a series of three more couplings in rapid succession to evaluate the effects of fatigue on his reflexes. Any fatigue resulting from the lengthy flight had not interfered with his ability to accomplish a successful coupling.

The thirty-fifth and final flight of the test program was flown on May 3. A movie crew aboard a camera chase plane filmed portions of the flight.

One hundred and seventy couplings had been accomplished during the tests. Twenty-four hours and eighteen minutes of towing time had been accumulated when the test project was concluded.

Wing tip coupling evaluation of the EC-47A and Q-14B continued at Wright Field for a year and a half after the conclusion of the initial tests. The team investigated the behavior of the coupled aircraft flying with one engine of the EC-47A shut down. An automatic roll stabilization system was tested and more pilots were familiarized with coupling the Q-14B to the EC-47A.

### *Single Engine Flight*

The performance of the coupled EC-47A and Q-l4B with one engine of the larger plane shut down was evaluated in the summer of 1950. At an altitude of 7,000 feet, Walt Rosenfield shut down the left engine of the EC-47A and feathered its propeller. The coupled airplanes flew on the power of the engine on the same wing as the Q-14B.

The rudder trim required for straight and level flight changed from 2.5 degrees nose left with both engines operating to 1.5 degrees nose right with the left engine windmilling.

Walt Rosenfield first banked the coupled pair to the right, toward the operating engine. Then he banked to the left, into the windmilling engine.

Rosenfield restarted the left engine of the EC-47A and shut down the right one. The dead engine and the Q-14B were both on the same side of the EC-47A. Nine degrees of nose left rudder trim was needed to cruise straight and level with the right engine shut down. Rosenfield repeated the evaluation of the coupled airplanes' behavior as he banked the airplanes to each side. It was easily possible to maintain altitude while flying at 130 mites per hour with one engine of the EC-47A operating near maximum power.

*The elevator autopilot control was slaved to a pair of micro switches on the rotating lance mechanism.(Air Force Film 14425)*

### *Automatic Roll Stabilization*

A primitive, automatic roll stabilization system was built into the coupling mechanism. A pair of micro switches was installed on the EC-47A receiver. A matching pair of actuating pins was mounted on the back of the lance mounting plate. Each pin deflected a micro switch as the Q-14B rolled away from the wings level position. The deflection of the micro switches served as the input to the elevator trim motor of the Q-14B to move the small plane back toward a flapping angle of zero. A toggle switch was installed in the cockpit of the Q-14B to turn the roll stabilization system on and off.

When engaged during its first test, the system moved the Q-14B to a high position relative to the desired attitude. The wing of the EC-47A was bending slightly in flight, causing the micro switches to be

*The microswitch controlled elevator autopilot hunted up and down unacceptably. (Air Force Film 14425)*

*The microswitch controlled elevator autopilot hunted up and down unacceptably. (Air Force Film 14425)*

*There were 230 couplings, fifty-six of which were made at night. (National Archives)*

activated at the wrong flapping angle. A series of trials were made to determine the proper position for the micro switches to return the Q-14B to the wings level attitude in a stable fashion.

The roll stabilization system produced a jerky oscillation of the Q-14B. The motor that operated the elevator trim tab moved the surface too fast. It tended to drive the Q-14B past the wings level position, so that activated the opposite micro switch. The airplane hunted up and down, never quite stabilizing at the proper position. It was not possible stabilize the Q-14B in hands-off flight even with the speed of the trim motor reduced by a factor of four.

The test team concluded that a rate-type roll stabilization system would produce superior results. Such a system would adjust the elevator trim based on the rate of change of the roll angle rather than the roll angle itself. Because the project was conducted with a philosophy of using simple systems to make a basic evaluation, it was decided not to make the major changes to the autopilot system that would be needed to introduce a rate-type system.

#### *Continued Evaluation, June 1950 - October 1951*

Evaluation of wing tip coupling continued with the EC-47A and Q-14B through 1951. The airplanes were assigned to the Headquarters of the Air Research and Development Command in April 1951 and then to the Wright Air Development Center in July 1951.

Thirteen more wing tip coupling flights were flown between May 3, 1950, and October 18, 1951. Another sixty-one couplings were accomplished, for a total of 231. The EC47A towed the Q-14B for another four hours and seventeen minutes.

Another eight pilots were familiarized with the technique, making a total of seventeen. Only three of the seventeen pilots were not able to engage the coupling ring with the lance.

### *Conclusions*

Bernard A. Hohmann of the Aircraft Laboratory at the Wright Air Development Center authored the report about the initial investigation of wing tip coupling. The team concluded that air-to-air wing tip coupling and flying were feasible. The exercise had provided valuable experience toward the further investigation of the method. It significantly bolstered the confidence of Air Force planners in the technique.

It was found that coupling could be accomplished with equal ease whether the approach was from the

the behavior of the coupled aircraft flying with one engine of the EC-47A shut down. An automatic roll stabilization system was tested and more pilots were familiarized with coupling the Q-14B to the EC-47A.

#### Single Engine Flight

The performance of the coupled EC-47A and Q-l4B with one engine of the larger plane shut down was evaluated in the summer of 1950. At an altitude of 7,000 feet, Walt Rosenfield shut down the left engine of the EC-47A and feathered its propeller. The coupled airplanes flew on the power of the engine on the same wing as the Q-14B.

The rudder trim required for straight and level flight changed from 2.5 degrees nose left with both engines operating to 1.5 degrees nose right with the left engine windmilling.

Walt Rosenfield first banked the coupled pair to the right, toward the operating engine. Then he banked to the left, into the windmilling engine.

Rosenfield restarted the left engine of the EC-47A and shut down the right one. The dead engine and the Q-14B were both on the same side of the EC-47A. Nine degrees of nose left rudder trim was needed to cruise straight and level with the right engine shut down. Rosenfield repeated the evaluation of the coupled airplanes' behavior as he banked the airplanes to each side. It was easily possible to maintain altitude while flying at 130 mites per hour with one engine of the EC-47A operating near maximum power.

#### Automatic Roll Stabilization

A primitive, automatic roll stabilization system was built into the coupling mechanism. A pair of micro switches was installed on the EC-47A receiver. A matching pair of actuating pins was mounted on the back of the lance mounting plate. Each pin deflected a micro switch as the Q-14B rolled away from the wings level position. The deflection of the micro switches served as the input to the elevator trim motor of the Q-14B to move the small plane back toward a flapping angle of zero. A toggle switch was installed in the cockpit of the Q-14B to turn the roll stabilization system on and off.

When engaged during its first test, the system moved the Q-14B to a high position relative to the desired attitude. The wing of the EC-47A was bending slightly in flight, causing the micro switches to be

*The elevator autopilot control was slaved to a pair of micro switches on the rotating lance mechanism.(Air Force Film 14425)*

*The microswitch controlled elevator autopilot hunted up and down unacceptably. (Air Force Film 14425)*

*The microswitch controlled elevator autopilot hunted up and down unacceptably. (Air Force Film 14425)*

*There were 230 couplings, fifty-six of which were made at night. (National Archives)*

activated at the wrong flapping angle. A series of trials were made to determine the proper position for the micro switches to return the Q-14B to the wings level attitude in a stable fashion.

The roll stabilization system produced a jerky oscillation of the Q-14B. The motor that operated the elevator trim tab moved the surface too fast. It tended to drive the Q-14B past the wings level position, so that activated the opposite micro switch. The airplane hunted up and down, never quite stabilizing at the proper position. It was not possible stabilize the Q-14B in hands-off flight even with the speed of the trim motor reduced by a factor of four.

The test team concluded that a rate-type roll stabilization system would produce superior results. Such a system would adjust the elevator trim based on the rate of change of the roll angle rather than the roll angle itself. Because the project was conducted with a philosophy of using simple systems to make a basic evaluation, it was decided not to make the major changes to the autopilot system that would be needed to introduce a rate-type system.

#### *Continued Evaluation, June 1950 - October 1951*

Evaluation of wing tip coupling continued with the EC-47A and Q-14B through 1951. The airplanes were assigned to the Headquarters of the Air Research and Development Command in April 1951 and then to the Wright Air Development Center in July 1951.

Thirteen more wing tip coupling flights were flown between May 3, 1950, and October 18, 1951. Another sixty-one couplings were accomplished, for a total of 231. The EC47A towed the Q-14B for another four hours and seventeen minutes.

Another eight pilots were familiarized with the technique, making a total of seventeen. Only three of the seventeen pilots were not able to engage the coupling ring with the lance.

### *Conclusions*

Bernard A. Hohmann of the Aircraft Laboratory at the Wright Air Development Center authored the report about the initial investigation of wing tip coupling. The team concluded that air-to-air wing tip coupling and flying were feasible. The exercise had provided valuable experience toward the further investigation of the method. It significantly bolstered the confidence of Air Force planners in the technique.

It was found that coupling could be accomplished with equal ease whether the approach was from the front or the rear. It was agreed that moving the lance forward into the coupling seemed more natural to the pilots.

The ailerons were ineffective at controlling the Q-14B while it was coupled to the EC-47A. Disconnecting the inboard aileron improved the roll control somewhat, but the ailerons still provided insufficient control authority. About the best thing that could be said was that flying with a single aileron did not increase the difficulty of coupling the airplanes.

It was considered advisable to install an adjustable trim tab on the outboard aileron of any aircraft with a disconnected inboard aileron. This would allow the pilot to trim the aileron so that it did not float up from the neutral position.

Each pilot typically required a few minutes to make his first coupling. With experience, the time required to accomplish the task was often reduced to less than twenty seconds. Major Anderson frequently performed the task in less than ten seconds.

It was necessary to conduct coupling operations at a speed well below the top speed of the Q-14B to allow an adequate margin of reserve power for maneuvering relative to the coupling ring. At high altitude, the Q-14B was flying too near to its maximum throttle setting. The pilot was left with very little throttle travel to control the speed of the airplane. Coupling operations should be conducted at the lowest altitude where smooth air could be found.

It was easier to couple the aircraft when the coupling ring was moved a short distance from the wing tip of the EC-47A. As a pilot gained experience, the task of coupling with the ring mounted directly to the wing tip became less daunting. The full-scale project design incorporated a telescoping boom in the wing of the B-29 to move the receptacle farther out from the wing tip. The boom would be retracted once the lance of the F-84 was locked in the receptacle.

Even slight turbulence made the task of engaging the coupling ring much more difficult. The pilot of the Q-14B had to insert the tip of his lance in an opening just four inches in diameter. Once the two aircraft were coupled, it was not difficult to fly the combination through rough air. On occasion, the degree of bending of the wing of the EC-47A in rough air was a source of consternation to the pilot of the Q-l4B. On three occasions the Q-14B was unintentionally released as the EC47A yawed suddenly in turbulence.

The pilot of the Q-14B could maintain the proper position while coupled on the wing tip of the EC-47A by reference to his instruments alone. The coupled pair could be flown in low visibility conditions that would have prevented two uncoupled airplanes from maintaining a close formation.

Single engine operation of the coupled aircraft presented no new problems. The EC-47A could be trimmed to fly on one engine with the Q-14B coupled to its wing tip.

Coupling the Q-14B to the wing tip of the EC-47A at night presented no special difficulties. The team recommended that lights provided for night operations should be equipped with rheostats to control their brightness. The lights should be shielded to restrict their illumination to the coupling mechanism on each wing tip. Areas of the upper surfaces of the airplanes that might be struck by the spotlights should be painted with dull, non-reflective paint to minimize unwanted reflections and glare.

The elevator became the primary control for maintaining wings level flight. Using the elevator, the pilot of the Q-l4B could follow the EC-47A through normal maneuvers. Turns, climbs, and descents presented no unusual problems to the pilots.

Using the elevator to control the roll of an airplane violates the natural reflexes of a pilot. It requires him to move his control stick at right angles to his natural inclination in order to achieve the desired flight attitude. As the right wing dropped, the pilot's nearly instinctive response would be to move the control stick to the left. The proper control input was to pull the stick back. Initially, flying the Q-14B coupled to the wing tip of the EC-47A required the full concentration of the pilot.

*The experience obtained was expected to further the development of a fighter and bomber combination. (Air Force Film 14425)*

After a short period of familiarization, most pilots became accustomed to the unusual method of control.

The elevator was more sensitive than was optimal for the task of maintaining wings level flight while couplets. It was felt desirable to decrease the sensitivity of the elevator in the full-scale development airplanes. This could be accomplished by the use of a gear reduction mechanism in the elevator or elevator trim system.

The ball joint coupling mechanism used for these tests allowed the Q-14B to rotate in pitch as the pilot applied the elevator to control the attitude of the airplane. The design of the coupling mechanism to be used in the full-scale project aircraft incorporated a rear latch to fix the wing tip of the F-84 in both pitch and yaw. Use of the elevator to control the attitude of the airplane would result in the application of torque to the wing tip of the B-29. The test team suggested that a floating rear latch should be investigated as a means of providing a limited freedom of movement in pitch for the F-84.

Pilots whose flying experience was primarily in large, multi-engine airplanes had difficulty making the transition to wing tip coupling with the small Q-14B. Their reflexes were conditioned to the slower response of the larger airplanes. They found it difficult to make the fine elevator control inputs necessary to successfully engage the lance in the coupling ring. Fighter pilots were found to have much less difficulty making the transition. Training in close formation flying was of primary importance to prepare a pilot for the task.

The basic roll stabilization system installed in the Q-14B was not successful. It was not possible to trim the Q-14B to fly without continuous pilot input. Until a suitable roll stabilization system could be perfected, long duration cruise flights would present a serious pilot fatigue problem. The longest coupled flight made by Major Anderson lasted 4 hours and 8 minutes.

It was recommended that the full-scale project development should be continued with a Boeing B-29 *Superfortress* and two Republic F-84 jet fighters. The basic methods utilized by this project did not allow for an evaluation of the range performance benefits of the wing tip coupled configuration. No attempt was made to seal the gap between the wings of the two airplanes to achieve the maximum benefit from the increase in aspect ratio. The full-scale development aircraft would incorporate seals between the wing tips of the B-29 and the F-84s.

It was suggested that the full-scale project should investigate the possibility of providing electric power, breathing oxygen, and fuel to the attached airplane through the wing tip coupling mechanism. Full-scale wing tip coupling tests with a B29 *Superfortress* and a pair of F-84Ds would begin at Republic Aircraft Company in Farmingdale, New York, in the summer of 1950.

#### *Subsequent Aircraft Assignments*

The designation of 44-68334 was changed to EQ-14B in September 1952, but it reverted to Q-14B in March 1953. It was relegated to class-32 status in April 1953. In January 1954 it was dropped from the inventory of the Air Force and transferred to museum status. It was subsequently carried on the U.S. civil register as N10146. In 1987 and 1988 it was owned by Morton W. Lester and was registered as N999ML. It is currently in the collection of the Experimental Aircraft Association Air Adventure Museum in Oshkosh, Wisconsin.

The flight test career of 42-23918 came to an end in December 1955. Its designation reverted to C-47A. It was transferred to the Military Air Transport Service in January 1956. For two years it hauled cargo for the 1739th Ferrying Squadron at Amarillo Air Force Base, Texas.

The next change of assignment put 42-23918 in the possession of 1405th Air Base Wing at Scott Air Force Base, Illinois, in February 1958. It made deployments to Keesler Air Force Base and Maxwell Air Force Base before it was transferred to the Warner Robins Air Logistics Area in Georgia in January 1961.

C-47A, 42-23918 was dropped from the inventory of the United States Air Force in February 1961. It was transferred to the Military Assistance Program and delivered to the air force of some other country.

### *References*

i Flight Investigation of a Wing Tip Coupling Method Using a C-47A and Q-14B Airplane; Bernhard A. Hohman; Aircraft laboratory, Wright Air Development Center; October 1951

ii To Fly and Fight; Col. Clarence E. "Bud" Anderson with Joseph P. Hamelin; Pacifica Press', 1990; Page 186.

iii To Fly and Fight; Col. Clarence E. "Bud" Anderson with Joseph P. Hamelin; Pacifica Press; 1990, Page 187.

*Republic EF-84D Thunderjets 48-0641 and 48-0661 coupled to the wing tips of Boeing EB-29A Superfortress 44-62093. (Bud Anderson)*

## Section 2: Project Tip Tow

### *Project Initiation*

Project Tip Tow investigated the feasibility of towing a pair of jet fighters attached to the wing tips of a bomber. Wing tip coupling held the promise of providing fighter escort for strategic bombers at a range far beyond that obtainable by jet fighters. It was suggested that a bomber could tow a pair of fighters attached to its wing tips. The bomber should be able to tow the fighters to the vicinity of the target with only a small reduction in its range. The fighters could be released to engage enemy fighter aircraft and then link up with the bomber to be towed home.

U.S.A.F. Contract 33-(038)-3202 authorized Republic Aviation Corporation to install wing tip coupling equipment in a Boeing B-29 Superfortress and two Republic F-84D Thunderjets and to conduct tests to determine the feasibility of the method.

The aspect ratio of the wing of the B-29 was 11.5. Coupling the Thunderjets to its wing tips increased the aspect ratio to 20.3 and improved the aerodynamic efficiency of the composite aircraft.

### *Range Extension of Bomber Escort*

In the post war period, the U. S. Air Force pursued a variety of methods of providing fighter escort for bombers on long range missions over enemy territory.

In 1948, the McDonnell XF-85 Goblin tested the feasibility of stowing a very small fighter internally in the bomb bay of a Convair B-36 bomber. The XF-85 was dead weight in the bomb bay until it was launched.

Wing tip coupling was envisioned as a method for a bomber to carry its own fighter escort with less impct on its range capability. Initial tests of the wing tip coupling concept were conducted by the Wright Air Development Center with a Douglas C-47 Skytrain and a Culver Q-14 in 1949. Sixteen pilots were trained to link the Q-14 to the wing tip of the C-47.

The success of the C-47/Q-14 wing tip coupling experiments prompted a plan to continue the investigation with a full scale bomber and apair of jet fighters. In 1950, a Boeing B-29 Superfortress and a pair of Republic F-84D Thunderjets were modified for a series of wing tip coupling test flights.

### *Aircraft*

Boeing B-29A Superfortress 44-62093 was selected for Project Tip Tow. It was delivered to the Army Air Corps on August 11, 1945. Its first assignment was the 241st Air Base Unit at Fairmont Field, Nebraska.

After the end of World War II, 44-62093 was relegated to training Strategic Air Command crews. Its designation was changed to TB-29A on June 13, 1946. It was transferred to the Air Defense Command and assigned to the 401st Air Base Unit at Hamilton Field, California in August 1946. It was returned to the Strategic Air Command and assigned to the 307th Bombardment (Very Heavy) Group at MacDill Air Field, Florida in October 1946.

Authority for 42-62093 was transferred to the Air Materiel Command in February 1947. It was assigned to the 4117th Air Base Unit at Robins Air Field, Georgia.

The Superfortress arrived at Wright-Patterson Air Force Base in March 1948. It was assigned to the 4000th

*The McDonnell XF-85Goblin investigated the concept of carrying a fighter inside the bomb bay of a bomber in 1948 and 1949. (McDonnell negative D4E 12155)*

Air Base Unit and deployed to Muroc in California. In February 1949 it was assigned to the 2759th Air Base Group at Muroc Field in California.

The designation of 44-62093 was changed to EB-29A on March 29, 1949. The "E" prefix indicated that an airplane had been modified for testing purposes. On April 19, its designation was changed to ETB-29A. Although it retained this designation for the rest of its career, the authors of the Republic Aviation Company report about the wing tip coupling tests referred to the airplane by the designation EB-29A. This history will follow their practice and use the EB-29A designation.

44-62093 was assigned to the 2750th Air Base Group at Wright-Patterson Air Force Base on May 20, 1949. On April 7, 1950, it was deployed to the Republic plant in Farmingdale, New York where it was modified for the wing tip coupling test program.

Two Republic F-84D Thunderjets, 48-641 and 48-661, were acquired from the 2750th Air Base Unit for the project.

F-84D, 48-661 was delivered to the Air Force on February 18, 1949. It was allocated to the Air Materiel Command. It was assigned to the 2750th Air Base Wing at Wright-Patterson Air Force Base. It was deployed to the Allison Engine Company in Indianapolis, Indiana one month later. It returned to the 2750th Air Base Wing in September 1949 and its designation was changed to EF-84D.

EF-84D, 48-661 was assigned to the MX-1016 wing tip coupling test project and transferred to Republic Aircraft Company in Farmingdale in November 1949.

F-84D, 48-641 was officially delivered to the Air Force on May 16, 1949, but it remained in the possession of Republic at Farmingdale until August 1950. Its designation was changed to EF-84D in April 1950.

### Personnel

Major C. E. "Bud" Anderson and Captain John Davis were selected to fly the Thunderjets. Anderson was the primary Q-14 pilot of the wing tip coupling tests conducted at Wright Field in 1949. Davis was the assistant chief of fighter operations and director of flight and all weather testing at the Wright Center. Captain Guy Townsend and Captain Russ Schleech were assigned to fly the EB-29A.

### Aircraft Modifications

Republic installed forward facing, wing tip coupling lances on the left wing of 48-661 and on the right wing of 48-641. The lances were oval in cross section. They had a limited range of rotation to accommodate the rotation of the Thunderjets around the wing tip of the Superfortress. The tip of each lance rotated 90 degrees to lock the lance into the receiver on the EB-29A. The lance installation added 120 pounds to the wing of the EF-84D.

*Boeing EB-29A Superfortress 44-62093 takes off from the Republic plant at Farmingdale, New York on July 1, 1950 for the first wing tip coupling test with an F-84. (Still frame from Air Force Film 18705: Range Extension of Bomber Escort)*

Republic constructed new outer wing panels for the EB-29A. Each contained a retractable boom with a cone-shaped receiver at the end. A hydraulic cylinder could extend the receiver about eighteen inches beyond the wingtip.

The interior of the receiver was oval in cross-section to conform to the lance on the wing tip of the EF-84D. The lance could not rotate within the receiver. The receiver was free to swivel from side to side to compensate for off-center contacts by the lance. It was spring-loaded to keep it centered.

The main boom cylinder was pivoted at the wingtip. A hydraulic cylinder that was attached to the inboard end of the boom acted as a snubber to absorb the energy of the forward motion of the fighter as it contacted the receiver.

A pitch lock mechanism was installed in the wing of the EB-29A to fix the angle of attack of the fighter relative to the bomber in coupled flight.

A large diameter roller was installed behind the lance on the wing tip of the EF-84D. It fit into a V-shaped bar on the wing tip of the EB-29A.

As the boom was retracted, the roller depressed a switch that actuated the pitch lock latch.

The latch swung out of the wing tip of the EB-29A, pushed open a spring-loaded door in the wing tip of the EF-84D, and then seated on the front end of the pitch lock pin inside the wing tip.

*Republic EF-84D Thunderjets 48-0641 and 48-0661 on the ramp at the Republic plant at Farmingdale, New York.(Air Force Film 18705)*

When the boom was locked in the retracted position, the Thunderjet was fixed to the wing tip in pitch and yaw.

Sponge rubber seals were placed on the wing tip of the EB-29A to fill the gap between the wing tips of the coupled airplanes.

A "LOCK - UNLOCK" switch was installed in the cockpit of the EF-84D to control the lance locking mechanism. Setting the switch to the Lock position activated a button on the wing next to the lance.

When the lance entered the receiver, it depressed the button and rotated the tip of the lance to secure it in place.

During the initial flight tests, the automatic locking mechanism was deactivated. The pilot controlled the rotation of the tip of the lance manually.

The rotation of the lance was hydraulically coupled to the ailerons. As the EF-84D rotated about the wing tip coupling, opposite aileron was applied to provide a restoring force. For each degree of rotation of the EF-84D around the coupling, the aileron was deflected 0.6 degrees. Aileron control inputs by the pilot of the EF-84D could override the lance-aileron interconnect system. The interconnect system could be easily disconnected between flights if desired.

A three position "Extend - Off - Retract" switch was installed on each side of the EB-29A cockpit to operate the hydraulic cylinder on that side of the airplane. The pilot operated the left boom and the co-pilot operated the right boom. A set of lights was provided for each pilot to indicate the status of several latches and hydraulic cutouts associated with the mechanism.

The coupling mechanism incorporated a system to jettison the EF-84D from the wing tip of the EB-29A in an emergency.

The original design of the jettison mechanism placed explosive squibs in the lance of the EF-84D. These squibs could separate the lance tip from the lance, allowing the EF-84D to move backward out of the socket and away from the EB-29A.

As a result of an incident involving difficulty separating the wing tip coupling Q-14 from the C-47 on February 24, 1950, the jettison mechanism was redesigned.

Explosive squibs were installed in the receiver fork retaining pins to separate the receiver from the EB-29A. The receiver would remain on the lance of the EF-84D as it pulled away.

The lance on the right wing tip of EF-84D Thunderjet 48-0641 fits into the retractable socket on the left wing tip of the EB-29A. (Air Force Film 18705)

The lance on the left wing tip of EF-84D Thunderjet 48-0661 (Air Force Film 18705)

The lance on the wing tip of the Thunderjet adds 120 pounds to the wieght of th eairplane.. (Air Force Film 18705)

A roller mounted at the rear spar of the F-84 seats in a v-fork on the B-29 wing. (Air Force Film 18705)

The crew of the EB-29A could fire the squibs to jettison the EF-84D only when the boom was in the retracted position. The circuit to detonate the squibs was broken as the boom was extended. Only the EF-84D pilot could jettison his airplane while the boom was extended.

The squibs were triggered automatically when the angle of the connection between the airplanes exceeded a pre-set limit. At the beginning of the test program, the flapping angle was limited to 15 degrees above or below wings level while the boom was in the extended position, and 10 degrees above or below wings level while the boom was retracted.

### Instrumentation

Photorecorders were installed in the test aircraft to record numerous flight parameters. The data recorded included control surface position, airplane attitude, elevator control force, yaw angle, and extension cylinder position. The right hand EF-84D was equipped with an indicator for the angle of the connection between it and the EB-29A.

*The pitch lock latch is hidden behind a spring loaded door. (Air Force Film 18705)*

*The button next to the base of the lance triggers the rotation of the locking tip when the lance is fully inserted in the socket. (Air Force Film 18705)*

*If the F-84 rolls toward or away from the B-29, a mechanical connection between the shaft of the lance and the aileron moves the controls  to conteract the roll. (Air Force Film 18705)*

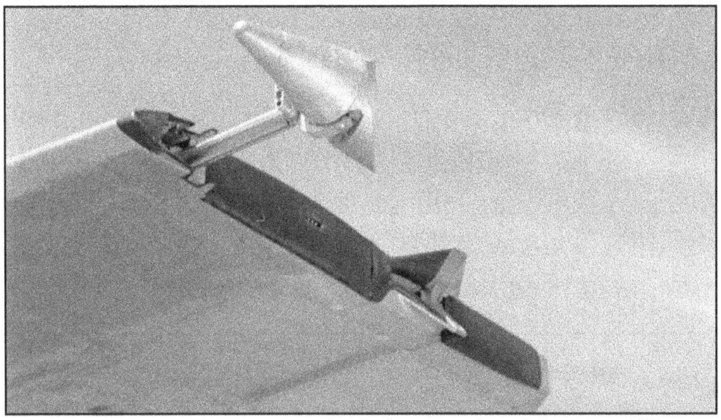

*The retractable socket on the left wing tip of Boeing EB-29A Superfortress 44-62093 extends nineteen inches from the wing tip. (Air Force Film 18705)*

*The socket can be hydraulically retracted to draw the wing tip of the F-84 flush against teh wing tip of the B-29. (Air Force Film 18705)*

*When the roller seats in the fork it actuates the rear latch within the B-29 wing. This locks the fighter to the bomber in pitch and yaw. (Air Force Film 18705)*

The cockpit signal lights that indicated the status of coupling related mechanisms were duplicated in the photorecorders. A system was provided to synchronize the photorecorders in the airplanes.

The wing of the EB-29A was instrumented to measure and record the loads and accelerations imposed on it.

A camera was installed in the aft fuselage of the left hand EF-84D to photograph checkerboard targets painted on the wing tips of the airplanes and on the fuselage of the EB-29A. Analysis of the images allowed determination of the amount of wing twisting imposed on the EB-29A by the EF-84D.

### Coupling Procedure

The pilot of the EF-84D maneuvered his airplane to insert the lance into the receiver on the wing tip of the EB-29A, much like probe and drogue refueling without the intervening cable. The pilot of the EB-29A flew straight and level to provide a stable platform.

*The modifications to the wing tips of the B-29 increased its weight by 700 pounds. (Air Force Film 18705)*

*A sponge rubber bumper formed a tight seal between the two airplanes.(Air Force Film 18705)*

The EF-84D pilot approached the bomber from astern, stabilizing in a position with the lance on his wing tip lined up with the receiver.

The EF-84D pilot applied slightly more power and guided the probe into the receiver. When the lance was properly seated in the receiver, he activated the lance locking mechanism with the "LOCK - UNLOCK" switch.

Once the EF-84D was locked in the receiver, the appropriate EB-29A pilot set his "Extend - Off - Retract" switch to the Retract position. As the boom retracted, limit and pressure switches cut out the snubber and locked the cylinder in the stowed position.

The pitch lock latch closed when the wing tip of the Thunderjet depressed a switch on the wing tip of the EB-29A. The status of each of these operations was indicated by a light on a console in the cockpit of the Superfortress.[i]

### Contractor Demonstration program

#### Ground Checks

Ground checks of the coupling mechanism were performed prior to the first flight of the wing tip coupling tests. A pair of overhead cranes lifted the Thunderjets off the ground. The receivers were extended from the wing tips of the EB-29A. The lances on the fighter wing tips were maneuvered into the receivers and locked in place. The receivers were retracted into the wing of the EB-29A and the function of the pitch locks was observed.

The EF-84Ds were rotated around the wing tip coupling. The team observed the operation of the interconnected ailerons. The Thunderjets were rolled far enough to trigger the jettison switches.

The sponge rubber seals between the wing tips were examined and adjusted to provide a good seal.

#### Functional Test Flights

The test aircraft were flown individually to evaluate the effect that their modifications had on their flight characteristics and to collect performance data. The EB-29A pilots detected little change in the behavior of the bomber with the receiver booms in either the retracted or extended position.

The weight of the wing tip lance caused the EF-84D to be slightly wing heavy. The EF-84D pilot could easily trim out the wing heaviness.

***July 21, 1950 Flight 1, First coupling***

Major Bud Anderson made first attempt to couple a Thunderjet to the wing of the Superfortress on July 21, 1950. He took off from Farmingdale in EF-84D, 48-661 with its lance mounted on its left wing. The lance-aileron connection was disconnected for the first test. A flat plate was installed in place of the pitch lock on the wing of the EB-29A. The plate was examined after the flight to determine where the roller of the EF-84D came into contact with it.

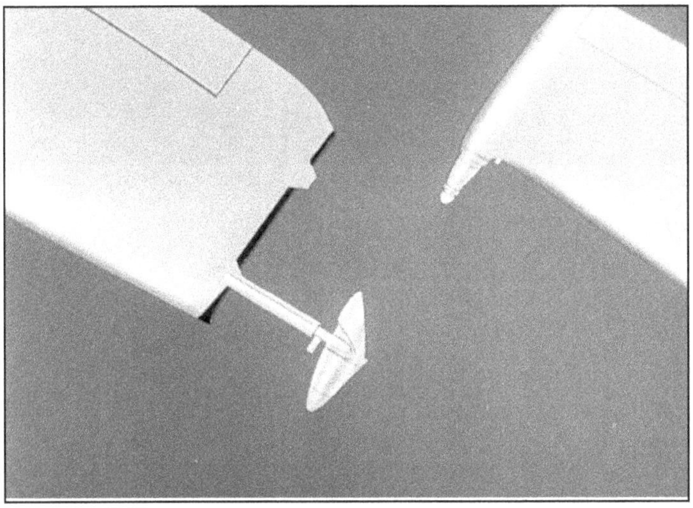

*The bomber extends its socket. (Air Force Film 18705)*

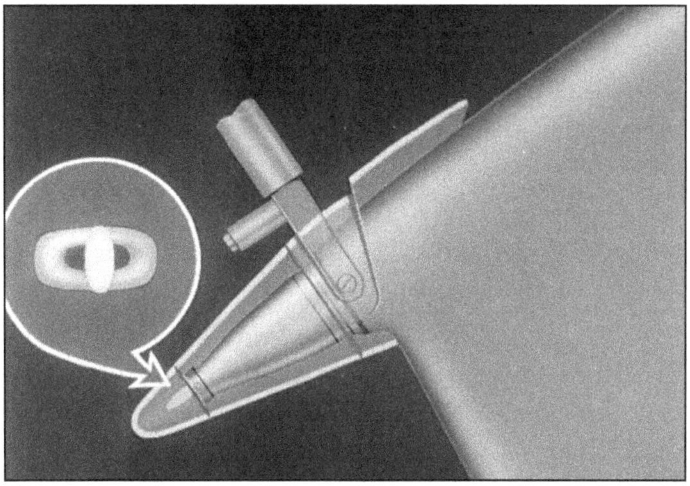

*The button actuates a switch, causing the tip of the lance to rotate 90 degrees. (Air Force Film 18705)*

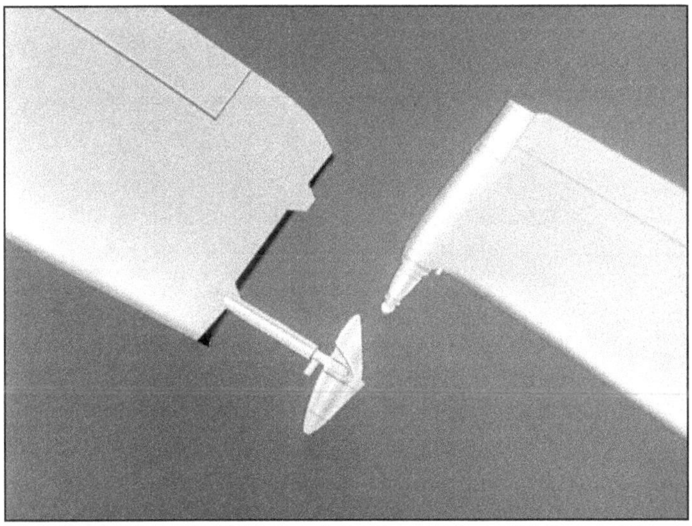

*The fighter inserts its lance into the socket.(Air Force Film 18705)*

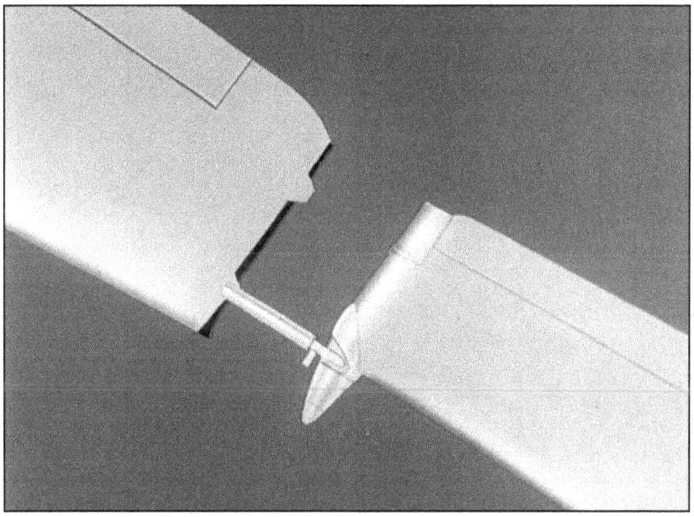

*Rotating the lance tip locks the fighter to the bomber in tow. (Air Force Film 18705)*

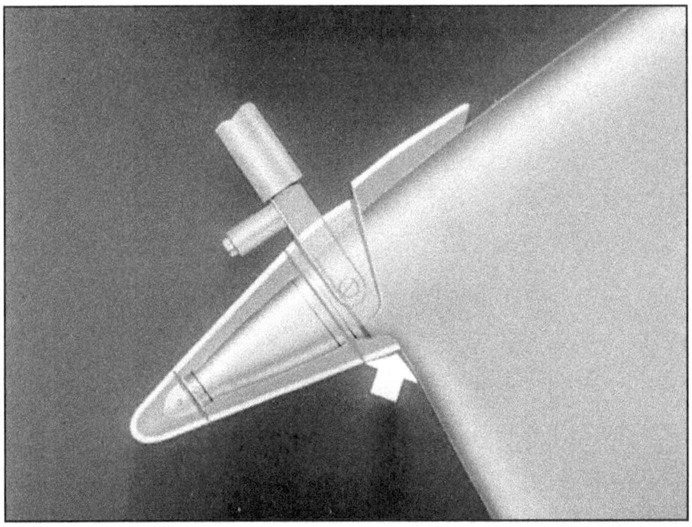

*The socket depresses a button on the leading edge of the fighter wing.(Air Force Film 18705)*

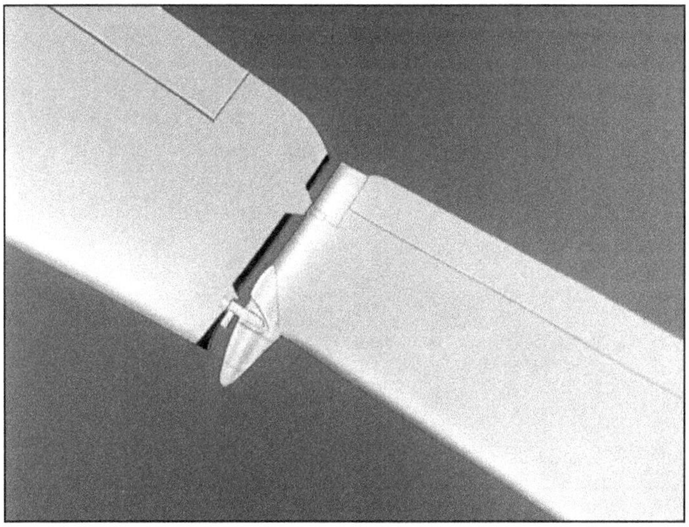

*The socket is retracted, drawing the fighter wing tightly against the bomber wing.(Air Force Film 18705)*

Captain John Davis flew chase on this first test flight. Davis would later fly the left hand Thunderjet in the wing tip coupling tests.

The performance characteristics of the EF-84D were not well matched to the EB-29A. The fighter was flying near its stall speed and the bomber was pouring on the coals to maintain 225 miles per hour. The lag in the throttle response of the jet engine was particularly pronounced at the low airspeed.

As with the initial efforts to couple the Q-14 to the wing tip of the C-47, Anderson found that the circulation around the wing tip of the EB-29A tended to force the lance up and away from the receiver.

The receiver was mounted on a swivel to allow it to turn toward the lance of the EF-84D when it was contacted off-center. Once in flight, Anderson discovered that the airflow around the wing tip of the EB-29A moved the receiver out of position.

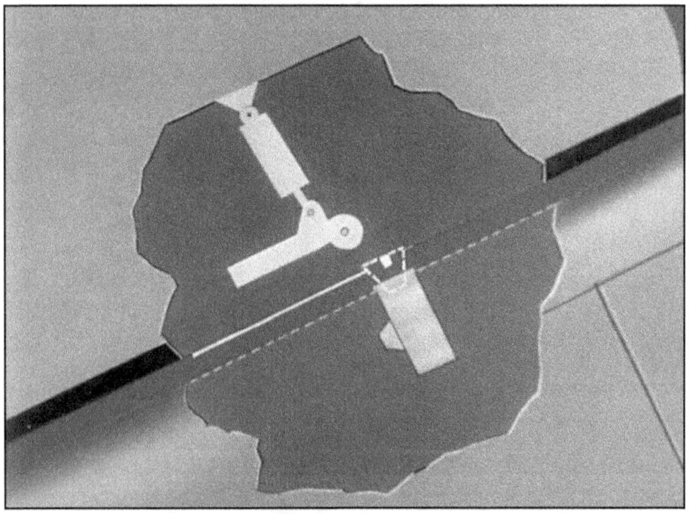

*The roller on the wing of the F-84 seats in the v-fork of the B-29. (Air Force Film 18705)*

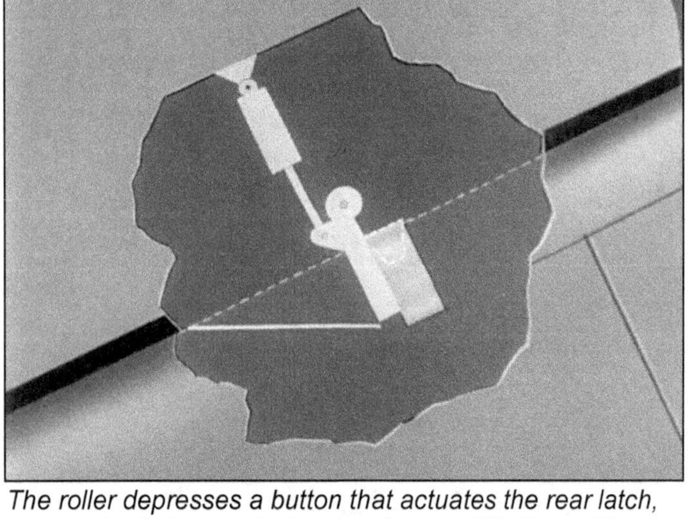

*The roller depresses a button that actuates the rear latch, locking the fighter to the bomber in yaw. (Air Force Film 18705)*

It required several approaches to the receiver for Anderson to insert the lance. He did not trigger the locking mechanism, but held the EF-84D in place with the thrust of the jet engine. After a short period of familiarization, he throttled back and let the lance slide back out of the receiver. He coupled his airplane to the wing tip of the EB-29A three more times.

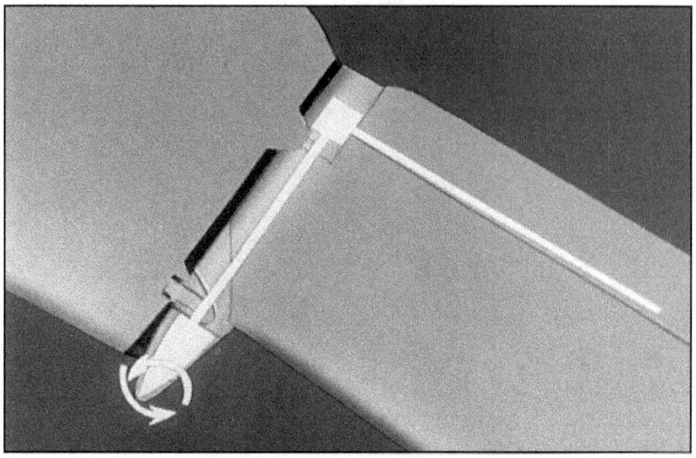

*A cam in the shaft of the lance allows the fighter pilot to override the mechanical inteconnect with the ailerons. (Air Force Film 18705)*

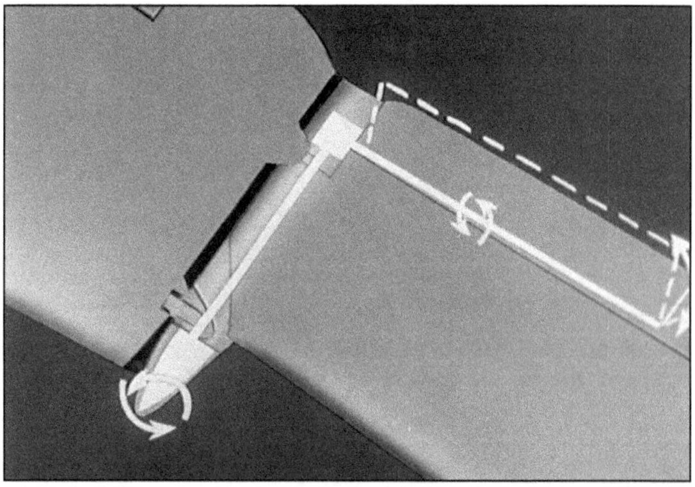

*As the F-84 rolls around the shaft of the lance, the ailerons move to counteract the roll.(Air Force Film 18705)*

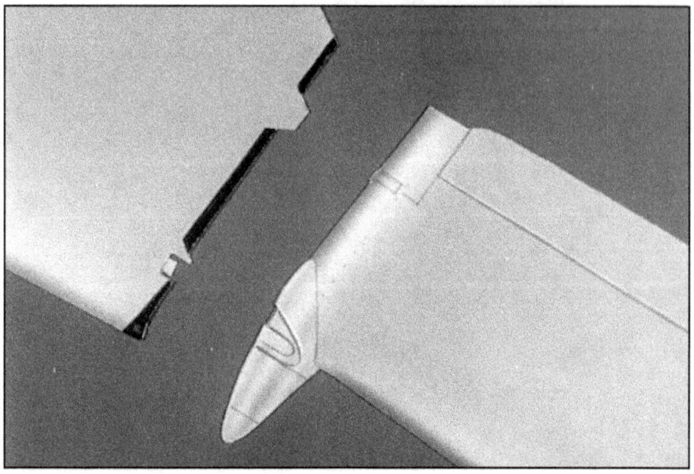

*The socket can be jettisoned from the bomber in an emergency. (Air Force Film 18705)*

Anderson discovered that wing of the EB-29A was much more flexible than the wing of the C-47. A steady approach at low closing speed produced a smooth coupling, but contacting the receiver with too much excess speed caused the wing of the Superfortress to begin flexing. Anderson compared the oscillating wing to a quivering diving board. The flexing increased in amplitude if left unchecked by the EF-84D pilot. Careless handling of the controls could lead to a pilot induced oscillation.

Anderson seated the lance in the receiver and triggered the tip lock. The co-pilot of the EB-29A initiated the retraction of the boom. The EF-84D remained stable as the boom was retracted, although the effectiveness of the ailerons diminished as the boom retracted. The changing lift distribution caused the right wing of the EF-84D to drop, which Anderson corrected by applying left aileron.

Once the boom was fully retracted, Anderson trimmed the EF-84D to fly level with the wing of the EB-29A. He found it easy to compensate for small movements away from wings level. He allowed the EF-84D to fly hands-off for a short period of time.

Anderson reduced the power of the jet engine of the EF-84D to idle. The drag of the fighter held it firmly against the wing tip of the EB-29A. Anderson found that the application of full left rudder was not sufficient to move the EF-84D away from the wing.

Anderson throttled up the engine of the EF-84D, and the boom was extended to the launch position. The Thunderjet remained steady during the extension.

Anderson disengaged the lance tip lock. He throttled back, and the lance slipped backward out of the receiver. Simultaneously, the latch pin slipped free of the pitch lock latch. Anderson had been attached to the bomber for seventeen minutes.

Following the flight, the flat plate that had replaced the aft pitch lock was examined for marks from the wing tip of the EF-84D. The marks were found slightly below the expected location, revealing that the EF-84D was flying at a higher angle of attack than had been anticipated. The misalignment might prevent proper locking of the rear latch on future flights. Additional fuel was loaded onto the EB-29A for the next flight to increase its angle of attack and improve the alignment of the wing tips of the airplanes.

#### *July 23, 1950 Flight 2*

A number of modifications were made to the coupling mechanism in preparation for the second flight. To reduce the binding of the lance in the receiver, the tolerances were opened up with a file. Sponge rubber bumpers were installed between the receiver and the aft face of the fork to keep it in the proper position.

The flexing of the wing of the EB-29A that followed a hard coupling was disconcerting to the flight crew. The crew of the Superfortress had no means to jettison the Thunderjet while the boom was extended. They were dependent on the pilot of the EF-84D to release himself from the coupling mechanism. They felt that the crew of the EB-29A should be able to jettison the Thunderjet at any time. An electrical cable was installed to allow the crew of the EB-29A to trigger the explosive squibs in the receiver the when the boom was extended.

The aft latch was installed in the wing tip of the EB-29A, but it was not activated. The interconnect between the lance and ailerons of the EF-84D was hooked up to investigate the effectiveness of the stabilizing system.

*Major Clarence "Bud" Anderson made the first coupling on July 1, 1950. (Air Force Film 18705)*

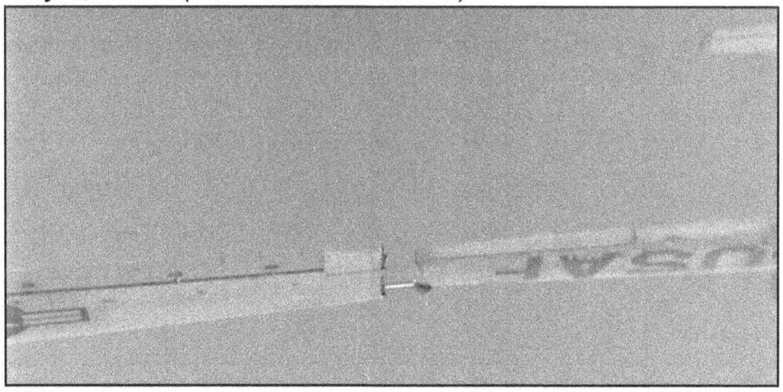

*Major Anderson made four successful couplings on the first day of tests. (Air Force Film 18705)*

*The average time to couple was two to ten minutes. (Air Force Film 18705)*

The second wing tip coupling test was conducted on July 23, two days after the first flight, Anderson found that the lance tended to stick in the receiver while only partially inserted. The lance rotated with every movement of the ailerons, so it did not stay aligned with the receiver. Part way in, the lance jammed in place. Anderson felt the control stick of the Thunderjet move from side to side as the EF-84D rolled around the wing tip coupling.

Anderson throttled back and separated from the receiver to try again. In spite of the additional sponge rubber padding, the receiver still swiveled objectionably when the lance hit it off-center. Eventually he found that he could seat the lance in the receiver by throttling up the jet as he made contact. He coupled the EF-84D to the EB-29A three times.

The coupled airplanes flew through turbulent air shortly after Anderson made the third coupling. The rough air caused the EF-84D to flap up and down on the wing tip of the EB-29A. The lance-aileron interconnect failed to reduce the flapping motion of the Thunderjet.

Anderson damped out the flapping of the EF-84D. A cam was incorporated into the lance-aileron interconnect mechanism to allow the pilot to override the controls. Post-flight examination revealed that control had been accomplished by twisting the shaft of the lance rather than the cam.

Once he had established stabilized flight on the extended boom, Anderson was able to trim the EF-84D to fly at whatever flapping angle he desired. The co-pilot of the EB-29A retracted the boom. The trim of the Thunderjet changed noticeably as the gap between the wing tips closed.

*Captain John Davis made the first coupling on the left wing of the B-29 on July 30, 1950. (Air Force Film 18705)*

*Captain Davis accomplished two couplings, despite a partial loss of aileron boost. (Air Force Film 18705)*

As the roller on the wing tip of the ED-84D engaged the pitch lock on the wing tip of the EB-29A it forced a small change in the pitch angle of the Thunderjet. That produced a change in the roll trim of the fighter.

The boom was retracted and Anderson reduced the throttle of the engine of the EF-84D to idle thrust. Anderson adjusted the elevator trim upward to compensate for the loss of engine thrust. Then he adjusted the aileron trim to compensate for the uneven lift distribution induced by the proximity to the wing tip of the EB-29A.

After a period of familiarization, the co-pilot of the EB-29A toggled the switch that extended the boom. The boom moved outward about four inches and then jammed. Although the boom had not reached the fully extended position, the green "Cylinder Extended" indicator lamp lit up. This was because the indicator lamp was triggered by the increase in hydraulic pressure that occurred when the boom stopped extending, whatever the actual position of the boom.

When it became clear that the boom was stuck, Anderson unlocked the tip of the lance and released the EF-84D from the receiver. Relieved of the drag of the EF-84D on the receiver, the boom then moved to the fully extended position.

Anderson's Thunderjet was coupled to the right wing tip of the Superfortress for nineteen minutes on the second test flight. He maneuvered to the other side of the EB-29A to evaluate the approach to its left wing tip. With no lance on his right wing tip, he was unable to hook up to the wing tip of the EB-29A.

### *July 27, 1950 Flight 3*

The coupling mechanisms were further modified for the third test flight. The pitch lock latch on the EB-29A was activated. The "Cylinder Extended" indicator lamp was re-wired to operate when the emergency release cable reached its full extension instead of being triggered by a rise in the hydraulic pressure of the cylinder. Additional sponge rubber bumpers were installed around the receiver fork to restrict the swiveling of the receiver.

The hydraulic system output was increased, and the cylinder extension pressure was increased to 1,000 pounds per square inch to improve the operation of the boom.

The maximum flapping angle of the EF-84D while in the retracted position was reduced from plus or minus 15 degrees. The emergency jettison would be triggered if the Thunderjet rolled 10 degrees above or 9 1/4 degrees below the wing tip of the EB-29A.

The third coupling flight was made on July 27. Captain John Davis flew chase in EF-84D, 48-641 with the lance on its right wing tip.

Restricting the swiveling of the receiver made it easier to insert the lance. Anderson was learning to anticipate the reaction of the EF-84D to the proximity of the EB-29A wing tip. With the proper application of power at the moment of insertion, he accomplished a relatively smooth hookup. He adjusted the trim of the Thunderjet, but induced a roll oscillation through small movements of the controls. He damped the flapping by centering the control stick.

The retraction of the boom jammed the receiver was about 4 inches from the wing tip. Anderson increased the throttle of the jet engine to 6,100 rpm and trimmed the controls of the EF-84D. Then the retraction was completed. The pitch lock engaged smoothly without producing as much of a trim change as on the earlier flights.

Anderson reduced power to seat the wing tip of the EF-84D in the pitch lock and secure the aft latch of the coupling mechanism for the first time. When he increased power to bring the engine speed up to 6,500 rpm, the aft latch yawed away from the wing tip of the EB-29A. He could not bring it back against the wing even with full application of the rudder of the EF-84D.

The EF-84D tended to drop below the wing of the EB-29A. The lance-aileron interconnect did not provide sufficient control authority to keep the Thunderjet level with the Superfortress. Despite application of corrective aileron, the EF-84D dropped down to the lower flapping angle limit, triggering the detonation of the explosive squibs that held the receiver on the boom. The receiver separated from the wing of the EB-29A. The EF-84D hung up on the aft latch momentarily until Anderson yawed the airplane away from the wing of the EB-29A. He had been coupled to the wing of the EB-29A for just seven minutes. Following the unintentional separation of the EF-84D, the extension cylinder could not be operated until the crew resorted to the emergency hydraulic system.

John Davis in EF-84D, 48-641 made the first coupling with the left wing tip of the EB-29A. He did not lock the lance into the receiver on that occasion. After thirty minutes of familiarization, he withdrew the lance and the airplanes returned to the Republic Plant.

***July 30, 1950 Flight 4***

In preparation for the next flight, the bolt that secured the roller in place was cut part way through, so it would shear and release the roller from the wing tip when the Thunderjet was jettisoned again. The automatic blow-off limit switches were set to plus or minus 15 degrees to reduce the chance that the EF-84D would be inadvertently jettisoned. The lance-aileron interconnect was disconnected.

The fourth wing tip coupling flight was made on July 30. Anderson had become proficient at threading the lance into the receiver and accomplished the task without incident. He trimmed the EF-84D to ride smoothly on the extended boom.

The co-pilot of the EB-29A toggled the switch to retract the boom, stopping the movement of the boom when the receiver was about 3 inches from the wing tip. Anderson throttled his jet engine down to idle thrust and adjusted the trim of the EF-84D before the boom was fully retracted. The aft latch failed to fully engage the lock pin when the EF-84D was drawn up against the wing tip of the EB-29A.

Anderson found that the aileron response of the EF-84D was relatively sluggish with the lance-aileron interconnect disabled. The interconnect had reduced the force required to apply corrective aileron.

Anderson left the jet engine at idle thrust as the boom was extended to the launching position. He re-trimmed the EF-84D and throttled up before unlocking the lance tip and separating from the EB-29A.

John Davis approached the EB-29A in EF-84D, 48-641 and inserted its lance into the receiver on the left wing tip. His control inputs caused the wing of the EB-29A to flex up and down, so he immediately withdrew the lance. He lined up for a second approach and successfully seated the lance in the receiver. While he was coupled to the wing tip of the EB-29A, small pieces were seen to fall away from the bomber. The

*Captain John Davis releasess his F-84 from the wing tip of the B-29. (Air Force Film 18705)*

test was immediately concluded. Post-flight inspection revealed that tape covering the gun ports of the turrets had torn away.

***August 8, 1950 Flight 5***

Before the fifth test flight on August 8, the retaining bolt for the aft latch roller was replaced with an explosive squib. Anderson succeeded in coupling his airplane to the EB-29A on the second attempt. He adjusted the trim of the Thunderjet while the boom was fully extended and again with the cylinder extended only three inches. He relied primarily on the elevator trim to maintain wings level flight. The engine of the EF-84D was throttled back to 6,100 rpm.

The rear latch secured the lock pin as soon as the wing tips came together. As the pitch lock engaged the roller of the EF-84D, trim changes caused the Thunderjet to roll to the right. Aileron trim was inadequate to keep the fighter wings level, so Anderson applied additional elevator trim.

After six minutes of coupled flight, the co-pilot extended the boom. Due to the changes in the elevator trim, the EF-84D pitched downward sharply as the pitch lock latch released. The remainder of Anderson's launch was conducted without incident.

Captain Davis inserted the lance of his EF-84D into the receiver on the opposite wing. He did not lock the lance in place. He separated from the EB-29A and made four more contacts with the coupling mechanism. His airplane was suffering a hydraulic leak in its aileron boost system, which increased the control forces. He averaged slightly less than one minute of coupled flight during each contact.

***August 15, 1950, transfer to Wright-Patterson Air Force Base***

Authority for the test airplanes was transferred to Wright-Patterson Air Force Base on August 15, 1950. The two Thunderjets were assigned to the 2750th Air Base Wing.

***August 22, 1950 Flight 6***

The lance-aileron interconnect in the EF-84Ds were reconnected. The sixth test flight was flown on August 22.

Anderson's objective on this flight was to compare the operation of the rear latch at full thrust and idle thrust. He made the initial coupling with no difficulty. The boom was partially retracted. Then Anderson throttled the engine down to idle thrust and adjusted the trim. The boom was fully retracted, but the rear latch did not lock.

The boom was partially extended again. Anderson throttled the jet up to a speed of 6,300 rpm and trimmed the controls. When the cylinder was retracted with power on, the rear latch locked properly.

The lance-aileron interconnect again proved inadequate. Anderson had to override the aileron control to keep the EF-84D properly aligned with the wing of the EB-29A

Once again, when the roller disengaged from the pitch lock, the EF-84D pitched downward, although not as severely as on the previous flight.

On Anderson's second attempt to couple the airplanes, the lance jammed when only partially inserted in the receiver. He kicked the rudder and waggled the ailerons in an attempt to free the lance unsuccessfully. He throttled back and pulled away from the coupling mechanism.

Captain Davis then approached the left wing tip of the EB-29A in the other EF-84D. He made five attempts to seat the lance in the receiver. Each time the lance jammed when only partially inserted. Several times his control inputs caused the wing of the EB-29A to start oscillating. Locking the control stick in the center position quickly damped out the oscillations. Rapid changes in EF-84D engine thrust caused the EB-29A to yaw noticeably.

Davis held the lance in place in the receiver with the thrust of his jet engine as the EB-29A made a shallow bank and turned 180 degrees. Anderson coupled his EF-84D to the opposite wing tip of the EB-29A and locked the lance in the receiver. The thrust of the two jet fighters increased the airspeed of the EB-29A by 10 miles per hour.

The lance of the Captain Davis' EF-84D was

*Both fighters were coupled to the bomber for the first time on August 22, 1950. (Air Force Film 18705)*

The composite aircraft in cruise configuration. (Air Force Film 18705)

examined following the flight. The locking tip was bent, which contributed to the jamming of the lance in the receiver.

### *August 24, 1950 Flight 7*

The next coupling attempt was flown two days later, on August 24. The rubber bumper at the aft end of the receiver fork was removed to allow the receiver greater freedom of movement.

Captain Davis' EF-84D was still grounded with a damaged lance lock. Major Anderson made four unsuccessful attempts to insert his airplane's lance into the receiver. Each time the lance jammed in the partially inserted position. No amount of rolling or yawing or the application of additional power could drive the lance home.

Analysis of the coupling mechanism suggested that when the lance contacted the receiver off center, the rotation of the receiver misaligned the components and caused binding. In preparation for the next flight, parts of the receiver were cut back and the sponge rubber bumpers were re-installed around the receiver fork.

The coupled airplanes were flown at altitudes from 1,000 feet to 20,000 feet. (Air Force Film 18705)

### *September 2, 1950 Flight 8*

The eighth flight of the test series was flown on September 2. Major Anderson coupled his EF-84D to the wing tip of the EB-29A first. The lance jammed, but Anderson applied more throttle and succeeded in seating the lance in the receiver. He engaged the tip lock.

Anderson reduced power to idle thrust and adjusted the trim while the boom was fully extended. He throttled up to 6,100 rpm, using the main flight controls to compensate for the trim changes. It was hoped that this would leave the trim adjusted for towed flight at idle thrust.

The co-pilot of the EB-29A retracted the boom. The aft latch did not seat properly on the pitch lock pin. Anderson manipulated the throttle and flight controls to secure the latch.

The pilot of the EB-29A rolled the coupled pair to the left and established a bank angle of 9 degrees. Anderson had to apply considerable force to the stick to keep the EF-84D at the appropriate bank angle. It required 9.5 degrees of aileron trim tab deflection to

The coupled airplanes were flown at airspeeds from 180 miles per hour to 230 miles per hour. (Air Force Film 18705)

trim out the stick force. Once properly trimmed, the EF-84D could be kept in position with little input from Anderson.

Instruments indicated that the wing of the EB-29A bent upward 8 degrees during the turn with the EF-84D. Wing loads were considered large, but not excessive.

Anderson throttled down to idle thrust. Only slight trim adjustment was necessary for short periods of hands-off flight. He throttled the engine up to 6,500 rpm prior to extension of the boom.

There was a slight adjustment in pitch as the pitch lock latch released the lock pin. The boom failed to extend fully. The EF-84D was released from the partially extended boom.

The EF-84D was still trimmed to keep the right wing up in towed flight. When the lance pulled free of the receiver, the Thunderjet rolled sharply toward the EB-29A. It nearly rolled upside down over the Superfortress' wing before Anderson regained control.

Captain Davis made four attempts to engage the receiver on the left wing of the Superfortress. Each time the lance jammed before it was fully inserted. Manipulation of the throttle and flight controls of the EF-84D failed to seat the lance.

Anderson had been coupled to the EB-29A for 21 minutes. Davis spent 30 minutes with his lance jammed in the receiver during his four attempts to lock it in place.

Davis and Anderson compared techniques during post-flight debriefing. Davis stated that he had to apply heavy aileron to maintain wings level flight as he approached the wing tip of the EB-29A. His application of aileron rotated the lance out of position, causing it to jam. Anderson related that he relaxed his restraint of the motion of the stick as the lance entered the receiver, allowing the lance to rotate to the correct position to seat in the receiver.

*Shallow turns, climbs, and let downs were acoomplished by the coupled trio. (Air Force Film 18705)*

***September 5, 1950 Flight 9***

The lance-aileron interconnect in Captain Davis' EF-84D was disconnected for the ninth flight. The pitch lock latch was installed in the left wing tip of the EB-29A, but it was not activated. The forward end of the receiver and the lance tips were filed down to improve clearance.

Captain Davis made the first attempt to couple his EF-84D to the wing tip of the EB-29A on September 5. With the lance-aileron interconnect disconnected, the lance was free to rotate as it was inserted in the receiver. Davis locked the lance in place on his second attempt.

The boom was partially retracted and Davis adjusted the. With the jet throttled up, the boom was retracted. Davis noted a small trim change as the roller seated in the pitch lock.

Davis' EF-84D was partially extended before Major Anderson made his approach to the right wing tip of the EB-29A, so that Major Davis could release himself quickly if a problem developed. Anderson made a particularly smooth engagement of the right wing tip receiver. His airplane was retracted to about six inches from the wing tip of the EB-29A. He trimmed the controls and then the booms were fully retracted. The three-airplane composite was fully assembled for the first time.

The behavior of each of the airplanes was affected by both of the others. If one EF-84D flapped downward, the EB-29A rolled in the opposite direction, causing the other EF-84D to flap upwards. The pilots easily compensated for these effects and had little difficulty

*Drag and stress were recorded undeer all test conditions. (Air Force Film 18705)*

maintaining the correct position relative to the wing of the EB-29A.

The pilot of the EB-29A banked the composite aircraft into a right turn. The pilots of the Thunderjets were able to follow the Superfortress into the turn, maintaining position during the turn and during the recovery from the turn.

The left hand EF-84D did not experience a significant change in pitch when the boom was extended because the pitch latch had not been engaged. The right hand EF-84D pitched sharply at the moment that it was released from the pitch lock. Once again, Anderson's EF-84D rolled sharply toward the EB-29A when he released the lance from the receiver. Anderson had spent nineteen minutes on the right wing tip, and Davis was attached to the left wing tip for 34 minutes.

An elevator trim tab position indicator was installed in the cockpit of each EF-84D. This allowed the pilot to set the elevator trim tab to the same position that it had been set to before the pitch lock had engaged, but the flight conditions at release did not exactly match those at the time of coupling. Engine power, flap setting, and fuel load all influenced the elevator trim setting.

***September 15, 1950 Flight 10***

The pitch lock latch in the left wing of the EB-29A was activated for the tenth test flight on September 15. It was the first attempt to secure the pitch lock in Captain Davis' EF-84D. An additional 18,000 pounds of gasoline was loaded into the EB-29A to increase its angle of attack. This improved the alignment of the wing tips during coupling.

Captain Davis seated the lance of the left hand EF-84D in the receiver without difficulty and locked the tip of the lance in place. The boom was retracted until the EF-84D was about six inches from the wing tip of the EB-29A. Davis reduced the power of the jet engine to idle thrust and adjusted the trim. He over-trimmed the aileron trim tab in anticipation of the trim changes that would occur during the last part of the retraction. When the retraction was completed, the pitch lock engaged immediately. Davis found that he had to apply left aileron to maintain the proper flapping angle, indicating that he had overcompensated when he trimmed the ailerons.

Davis' EF-84D was extended a short distance from the wing tip of the EB-29A to permit Anderson to attach his EF-84D to the other wing tip. The pilot of the EB-29A rolled the three airplanes into a shallow bank while the Thunderjets were still partially extended. Uncoordinated control inputs or power changes caused the two EF-84Ds to oscillate in pitch and yaw slightly. The pilots easily damped out the oscillations.

The pilots of the Thunderjets throttled back to idle thrust and trimmed their controls in preparation for full retraction. Both Thunderjets experienced a slight trim change as the aft locks engaged. It required a few moments of control adjustment to get the pitch latches to lock on the pin of each EF-84D. This was the first completely locked, towed flight of both Thunderjets at the same time.

The pilots of the Thunderjets trimmed their controls so that each EF-84D was banked slightly toward the EB-29A. This was a more stable flight attitude for the fighters. The pilots of the Thunderjets were able to fly hands-off for relatively long periods of time.

With the three airplanes in the cruise configuration, the collection of flight performance data could begin. The power of the EB-29A was adjusted to a series of different settings and the airspeed at which it stabilized at each power setting was noted.

Performance data for each of the airplanes had been collected individually. Speed and power data was collected for the EB-29A flying by itself..

*The engines of both fighters were successfully shut down for several minutes at a time. (Air Force Film 18705)*

*Successful air starts were made of the F-84 engines at all altitudes. (Air Force Film 18705)*

The weight added to the EB-29A improved the alignment of the wing tips. The pitch change as the aft locks disengaged was much less pronounced than it had been previously. With their engines at idle thrust, the fighters yawed slightly away from the EB-29A as the booms were extended.

Anderson and Davis released the lances of their airplanes from the receivers of the EB-29A. As Anderson's EF-84D separated, it tipped the receiver out of position. It would not return to the center position. Anderson pushed the receiver back into the proper position with the lance on his airplane.

The Thunderjet pilots engaged the receivers of the EB-29A again. They were partially retracted to throttle down their engines. In order to reduce the yawing induced by asymmetric changes in thrust, the pilots of the Thunderjets synchronized the reduction in power to idle thrust. The Thunderjet pilots shut off their jet engines at the same time and let the turbines windmill. They noted little change in the trim of their airplanes other than a slight yaw away from the EB-29A as the power was shut down.

The speed of the retraction of the boom was considerably slower with the engines of the EF-84Ds windmilling. The extra drag caused the cylinders to bind. The last six inches of retraction required nearly five minutes. The "Aft Lock Latched" indicator lights in the EB-29A cockpit illuminated as soon as the retraction was complete. The outward yaw of the EF-84Ds caused the pitch lock latches to seat immediately.

Towed flight with the jet engines of the Thunderjets windmilling was much the same as towed flight at idle thrust. The team collected additional speed and power performance data with the jets shut down.

The pilots of the Thunderjets air started their engines while the booms were still fully retracted. The jet ignition switch was on the right side of the EF-84D cockpit, and the pilot had to take his right hand off of the control stick to reach it. This required him to remove his left hand from the throttle to grab the control stick for a moment.

The booms were extended, and the pilots of the Thunderjets released their lance tip locks to separate from the EB-29A. The lance of Anderson's EF-84D tipped the receiver away from center again.

Anderson had been towed for 34 minutes and Davis was towed for 45 minutes. Thirty minutes were flown with both Thunderjets being towed by the EB-29A.

The left aileron of Anderson's EF-84D was disconnected for the next flight. When the Thunderjet was coupled to the wing tip of the EB-29A, the inboard aileron moved in the wrong direction for controlling the flapping angle. Disabling the inboard aileron should improve aileron control authority in coupled flight. The downside was that it reduced aileron control effectiveness in normal flight.

Major Anderson conducted a functional and flight characteristics check flight of the single aileron configuration of his airplane. He determined that aileron control was satisfactory for take-off and flight at speeds up to 300 miles per hour. Although the rate of roll was somewhat reduced, there was still adequate control.

The aileron trim tab of the EF-84D type airplane is mounted on the left aileron. With the left aileron fixed in place, Anderson had no means to adjust the aileron trim of his airplane as flight conditions changed. A large, fixed trim tab was installed on the right aileron. The trim tab was adjusted on the ground to produce zero aileron force with the stick centered while flying at 225 miles per hour in undisturbed air. Any other flight conditions would require Anderson to apply some amount of stick force to maintain wings level flight.

*The engines of the F-84s were normally left idling at 18% power while recording the total drag of the coupled airplanes. (Air Force Film 18705)*

*Over ninety couplings were accomplished during the fifteen test flights. (Air Force Film 18705)*

Anderson established level flight at a relatively high altitude and throttled his engine back to idle thrust. He dropped the nose of the EF-84D until it was diving at the airspeed at which it was towed by the EB-29A and noted the rate of descent. He throttled up and climbed back to altitude. Then he shut off the fuel flow to the engine and let the turbine windmill as he conducted another check of the rate of descent at towing speed. The rate of descent checks provided baseline drag data for the EF-84D to be used in the analysis of the performance data collected by the composite aircraft.

Anderson evaluated the hydraulic pressure and flight control effectiveness with the engine shut off. In upcoming tests, the engine of the EF-84D would be shut down during towed flight. The rotation of the turbine in the engine generated the hydraulic pressure used to operate the flight controls. Anderson established that the windmilling of the turbine generated adequate hydraulic pressure to control the Thunderjet.

***October 13, 1950 Flight 11***

Nearly a month passed before the eleventh wing tip coupling flight, which was flown on October 13. Only Anderson coupled his EF-84D to the EB-29A that day. The lance-aileron interconnect of 48-661 was disconnected.

Flying a steady approach to the receiver with only one functioning aileron was more difficult. After Anderson had locked the lance in the receiver, the boom was partially retracted. He throttled the jet engine back to idle thrust and the boom was fully retracted. The "Rear Lock Latched" indicator light in the cockpit of the EB-29A did not go on when the boom reached the fully retracted position. Anderson kicked in a little right rudder to seat the pitch lock latch and illuminate the indicator light.

*One night flight was made, during which four couplings were completed on each side.(Air Force Film 18705)*

Once coupled, the rate of roll of the Thunderjet about the wing tip of the EB-29A with the single aileron was improved, but it was still not sufficient to maintain the proper flapping angle in turbulent air.

The airspeed of the coupled pair was slowly reduced from 225 miles per hour to 200 miles per hour to measure the twisting loads imposed on the wing of the EB-29A by the EF-84D. As the airspeed of the Superfortress was reduced, the Anderson had to apply more up elevator to adjust the trim of the Thunderjet. Using the elevator to adjust the trim of the EF-84D imposed a twisting force on the wing of the EB-29A.

Strain measurements recorded during the flight indicated that the twisting force, while significant, was not likely to damage the structure of the wing of the EB-29A. This test confirmed that the elevator was the appropriate primary control for wingtip-coupled flight. It was suggested that the twisting force could be reduced by adding more weight to the EB-29A to increase its angle of attack.

The flapping angle of the EF-84D increased as the airspeed of the EB-29A was reduced. The angle of attack of the Superfortress increased as the airspeed was reduced. Because the EF-84D was fixed to the wing tip of the EB-29A in pitch, its angle of attack increased as well. The extra lift generated at the increased angle of attack caused the Thunderjet to stabilize at a higher flapping angle. The extra lift also caused the wing of the EB-29A to bend upward.

At the conclusion of the low speed test, the pilot of the EB-29A increased the airspeed to 225 miles per hour. Anderson throttled up the jet, and the co-pilot of the EB-29A toggled the switch to extend the boom. The boom failed to extend freely. Anderson throttled up the jet engine of the EF-84D, but the cylinder moved slowly. Eventually, the boom extended far enough for Anderson to release the lance of the EF-84D from the

*Tests indicated that a pair of F-84s could be towed by a B-29 carrying a 10,000-pound bomb load with a reduction in range of only 2.9%.(Air Force Film 18705)*

receiver. The Superfortress towed the Thunderjet for 35 minutes.

***October 19, 1950 Flight 12***

The project moved from Farmingdale, New York to Wright Field in Ohio for collection of performance data on long endurance missions.

The right aileron of Captain Davis' EF-84D was disconnected for the twelfth flight of the project, which was flown on October 19. Unlike Anderson's Thunderjet, 48-641 retained the function of the aileron trim tab mounted on the left aileron, so it was possible for Davis to trim out aileron stick forces as flight conditions changed. An additional 2,000 pounds of fuel was loaded onto the EB-29A to increase its angle of attack.

Major Davis coupled his EF-84D to the EB-29A first. Full application of the available aileron trim did not keep the Thunderjet level with the wing tip of the EB-29A. Davis had to apply constant force to the control stick to maintain wings level flight as he maneuvered the lance into the receiver. Anderson then coupled his EF-84D to the other wing tip. The aircraft trim was adjusted, and the engines were throttled down. The booms were partially retracted and the Thunderjets were re-trimmed. Then they were fully retracted and elevator trim was adjusted for level flight. The "Aft Lock Latched" indicator lamps failed to illuminate when the booms were retracted.

Initial tests were conducted at 20,000 feet altitude. Turns at bank angles of up to 10 degrees were made to both the left and right. The EF-84D pilots had to apply considerable aileron force to maintain position in the turns.

The pilot of the EB-29A stabilized in straight and level flight at a variety of power settings. Airspeed and fuel flow were noted at each power setting for comparison to data collected by the EB-29A flying on its own. Data was collected at airspeeds ranging from 225 miles per hour down to 185 miles per hour. A North American F-51D Mustang operated by the Air Materiel Command paced the composite aircraft.

*The effectiveness of the lance/aileron interconnect was marginal. (Air Force Film 18705)*

The control of the Thunderjets remained good at the lower airspeeds. Application of elevator easily compensated for the small change in trim.

The booms were partially extended for a descent to 15,000 feet altitude, where the next set of data values was to be collected. The Thunderjets pitched and yawed around the receiver during the descent. It required the constant attention of the pilots to damp out the oscillations.

The "Aft Lock Latched" indicator lamps illuminated when the booms were retracted at 15,000 feet. The pilot of the EB-29A ran through another series of power settings for speed and fuel flow data collection.

Another sequence of boom extension, letdown and boom retraction brought the test altitude to 10,000 feet. After the pilot of the EB-29A ran through the usual sequence of power settings the airspeed was increased to 225 miles per hour to release the Thunderjets. Anderson's EF-84D rolled to the right when it was released from the receiver. The lance tipped the receiver upward where it remained out of position.

Davis neglected to re-trim the elevator of his EF-84D after throttling up his jet engine. As the pitch lock disengaged, his Thunderjet pitched up and rolled sharply toward the EB-29A. After he stabilized the EF-84D, he made an otherwise normal release.

Anderson had been coupled to the right wing tip for one hour and thirty-nine minutes. Davis was coupled to the left wing tip for one hour and forty-seven minutes. After the flight, it was discovered that the photorecorder camera in the EB-29A had malfunctioned. None of the data had been recorded. It was necessary to repeat the flight the following day.

***October 20, 1950 Flight 13***[ii]

On October 20, another long endurance data recording flight was flown to get the data that had failed to be recorded the day before. Anderson and Davis had become fairly proficient at locking the lance in the receiver. They climbed to 20,000 feet altitude, rendezvoused with the EB-29A, and coupled to its wing tips in just fifteen minutes.

The composite aircraft made turns in both directions. The EF-84D pilots found that heavy aileron forces were required during the turns even when the airplanes had been trimmed for zero control force just before the turn was begun.

Speed and power data were collected at specified power settings. Data was collected at 20,000 feet, 15,000 feet, and 10,000 feet. When a right turn was initiated at 10,000 feet altitude, the control forces in the Thunderjets were so high that the pilots called for a halt to the turn so that they could readjust the trim settings.

Anderson's EF-84D was coupled to the wing of the EB-29A for two hours and thirty-seven minutes on that flight. Davis accomplished the longest wingtip-coupled flight of the program. He stayed coupled to the EB-29A for two hours and forty-four minutes.

Flying with the fighters attached to the wing tips changed the trim characteristics of the bomber. The EB-29A flying solo required aileron trim adjustments in response to changes in engine thrust and flight loads. Those trim changes were considerably reduced when the Thunderjets were attached to the wing tips. The EB-29A by itself flew with right rudder. The amount of trim required varied with airspeed. With the Thunderjets attached, the composite aircraft flew with left rudder trim that did not vary with airspeed.

The pitch stability of the EB-29A was reduced slightly by the presence of the Thunderjets. The pilot of the EB-29A easily accommodated the small reduction in pitch stability.

### Analysis

The thirteenth flight of the program concluded the contracted evaluation of the EB-29A/F-84D composite aircraft. Anderson and Davis had accomplished 43 in-flight couplings. The two EF-84D Thunderjets had accumulated a total of fourteen hours and fifty minutes of wing tip coupled flight.

The basic feasibility of coupling an EF-84D to the wing tip of a EB-29A was demonstrated. The fighter pilots were able to learn the proper method for engaging the receiver on the wing tip and to control the attitude of the fighter while coupled to the EB-29A. Disengaging the EF-84D from the receiver was a simple matter.

The receiver booms were left in the extended position during early flights. Anderson and Davis did not like the behavior of the fighters when they were attached to the extended booms. When the bomber performed anything more than gentle, banking maneuvers, the fighters would start wobbling around their attachment points and the wing of the bomber would flex like a diving board.

The composite aircraft was controllable in turns, climbs, and descents, although the fighters oscillated unpleasantly around the wing tip coupling mechanisms during descents with the booms extended. The pitch stability of the composite aircraft was slightly reduced compared to a standard EB-29A, but the EB-29A pilot could easily compensate for the reduction in stability.

The team demonstrated that the composite aircraft could be flown with the engines of the Thunderjets shut down. Starting the engines of the EF-84Ds while attached to the EB-29A posed no problem.

Disconnecting the inboard aileron of the fighters improved the aileron control authority in coupled flight, but it reduced the rate of roll and controllability in separate flight. This was particularly problematic while inserting the lance in the receiver.

Connecting the operation of the ailerons to the rotation of the lance improved the lateral control of the Thunderjets while coupled to the Superfortress. Limited periods of hands-off flight could be flown, but the system could not keep the EF-84D in the correct position during maneuvering or in turbulence.

After demonstrating that the ailerons did not provide enough control authority to do the job, the test team disconnected the lance-aileron interconnect for the remainder of the flight tests.

The elevator proved more effective than the ailerons for trimming the Thunderjets for flight coupled with the Superfortress. The application of elevator had the undesirable side effect of twisting the wing of the bomber. It was determined that the wing structure was strong enough to absorb the load of the twisting force.

*It was felt that a hydraulic control system using only the outboard aileron would provide superior control authority. (Air Force Film 18705)*

Dr. Vogt's theories regarding the range benefits of the increased aspect ratio of the composite aircraft were supported by the test results. Analysis of the performance data collected on the last flight confirmed that the drag of the composite aircraft was less than that of the three airplanes flying independently under most conditions. The method of wing tip coupling offered a potential range advantage over towing the Thunderjets behind the EB-29A on cables.

The reduction in drag achieved by wing tip coupling was most pronounced at low airspeed and high gross weight. At lower airspeeds, the drag of the composite aircraft was actually less than that of the EB-29A flying alone. Guy Townsend and Russ Schleech, the pilots of the EB-29A, reported that the bomber's speed actually increased slightly when the fighters had hooked up and shut down their engines.

At higher airspeed, the pilot of the EB-29A had to add power to offset the drag of the EF-84Ds. In cruising flight, the composite aircraft experienced about a three-percent reduction in range compared to the EB-29A flying solo. Under these conditions, towing the fighters on cables would require less power than wing tip coupled flight.

The flapping angle of the fighters varied with the airspeed of the composite airplane. The design value for the flapping angle was only obtained at a single angle of attack. The EF-84Ds were banked toward the EB-29A under most conditions. The constant bank angle was a source of discomfort to the fighter pilots.

The actual drag reduction achieved during the flight tests was less than the theoretical prediction. The twisting of the wing of the EB-29A and the divergence of the flapping angle of the Thunderjets from the design value impacted the drag reduction of the composite aircraft.

The glide ratio of the composite aircraft was increased relative to the airplanes operating separately. The distance traveled during let down to a landing field was longer for the composite aircraft.

The test aircraft were not equipped to transfer fuel from the EB-29A to the Thunderjets. The fighters were limited to the fuel that they carried at take-off. Any fuel that they consumed prior to hooking on to the EB-29A was not available for maneuvering in the combat area and landing.

To conserve the fuel of the EF-84Ds, they were hooked to the EB-29A at 5,000 feet altitude. Their engines were shut down, and the rest of the climb was conducted under the power of the Superfortress. The drag of the EF-84Ds with their engines windmilling significantly reduced the rate of climb and increased the fuel consumption of the EB-29A during the climb to altitude. The amount of fuel burned by the Superfortress in the climb could be reduced by utilizing the thrust of the jet engines of the EF-84Ds, but the cost in jet fuel was too great to permit use of the method except in emergency conditions.

### Conclusions

The method of wing tip coupling held the promise of providing fighter escort for strategic bombers at a range far beyond that obtainable by the jet fighters of the day. This escort could be carried along with only a small reduction in the range of the bomber. It was expected that the system to be deployed would use B-50 Superfortresses as the carrier aircraft.

The contractor's test report concluded that, "in-flight, wing tip towing of fighter aircraft with the equipment installed in the test airplanes and under calm air conditions is considered to be sufficiently promising to warrant further investigation and improvement of the towing systems."

### Recommendations

It was felt that the method of wing tip coupling would benefit from an adjustable rear support for the fighter. An adjustable support would alleviate the twisting force on imposed the wing of the EB-29A. It would also allow the EF-84Ds to be trimmed to fly wings level at any airspeed while coupled to the Superfortress.

It was felt advisable to strengthen the wing of the EB-29A to reduce the twisting caused by the Thunderjets on the wing tips.

To maximize aileron control in all flight conditions, the team recommended investigating a means to cut out the inboard aileron during coupled flight, while providing full aileron control for the EF-84D during independent flight.

The team recommended that an elevator stick lock should be installed to fix the control in place once the elevator had been trimmed for coupled flight.

The team suggested that installing a means to shut off the inlet to the engine would substantially reduce the drag of the windmilling engines.

### Range Missions

A variety of range missions were calculated to

determine the most advantageous method for utilizing the composite aircraft in combat.

A baseline mission was calculated for a EB-29A flying on its own at 10,000 feet altitude. The baseline mission formed the basis for evaluation of the range benefit of wing tip coupling. The take-off weight of the EB-29A was 135,000 pounds. It carried 6,000 gallons of gasoline and 17,600 pounds of bombs. An 800-gallon reserve of fuel was maintained for landing.

The Thunderjets weighed 14,195 pounds at take-off. They carried 410 gallons of fuel and 450 pounds of ammunition. 45 gallons of fuel were held in reserve for landing.

A range mission was calculated for the composite aircraft flying a mission profile similar to the EB-29A operating independently. The range of the composite aircraft was reduced approximately 230 miles. This represented a range reduction of 7.5%.

The profile of one composite aircraft mission involved coupling the fighters to the bomber at an altitude of 5,000 feet. The composite aircraft then climbed to 15,000 feet for the cruise to the target area. Before reaching the target, the composite aircraft climbed to an altitude of 20,000 feet for weapon release and the cruise back to base. The composite aircraft consumed more fuel in the climbs and fuel mileage at the higher altitudes was not as good, so the range was reduced in comparison to the lower altitude mission.

Eliminating the climb to 20,000 feet and the high altitude cruise back to base resulted in a mission range only about 138 miles less than the EB-29A operating alone, which was a 4.5% reduction in range.

The total range of the mission could be increased by flying the fighters to cruise altitude separately before coupling them to the wing tips of the bomber. The reduction in fuel remaining available to the fighters limited them to 20 minutes of combat at the target. The reduction in the fuel consumption by the EB-29A during the climb to the cruise altitude resulted in a range that was just 89 miles less than that of the EB-29A by itself, a reduction of only 2.9 percent of its normal range.

The team evaluated the benefit of providing a means to refuel the EF-84D from the EB-29A. By reducing the bomb load of the EB-29A to provide 800 gallons of jet fuel for the fighters, the range of the composite aircraft was about 140 miles less than that of the EB-29A flying separately, assuming that the fighters operated independently for only 20 minutes in the target area.

### *MX-1152 - Long Line Tow*

The Air Force conducted other fighter escort range extension tests in competition with Project Tip Tow. Cable towing tests were conducted at Wright-Patterson AFB and Edwards AFB under the auspices of the MX-1152 - Long Line Tow project in the spring of 1951. Boeing B-50 Superfortress, 49-0307 was equipped with a cable and winch to test the practicality of towing a fighter behind the bomber. Lockheed T-33A, 49-0938 was equipped with a towline attachment point on its nose.

*MX-1152 Long Line Tow investigated towing Lockheed T-33A 49-0938 on a cable behind Boeing B-50D Superfortress 49-0307 at Edwards AFB in the spring of 1951. (Air Force Flight Test Center history Office)*

Captain Stener of Wright-Patterson piloted the B-50 tow plane. Captain Newman flew the T-33A. Captain Hunt served as the photo-chase pilot in a Lockheed RF-80. Major Neil N. Lathrop was assigned as the Edwards AFB Project Officer. Eleven flights were conducted between February 22 and March 8.

The B-50D arrived at Edwards Air Force Base with the crew and technicians for the project on February 20, 1951. The B-50D pilot for the project was Captain Stener. The pilot of the T-33 was Captain Newman. Captain Hunt piloted a Lockheed RF-80 as the photo chase plane. The jets arrived at Edwards Air Force Base on February 22, 1951.[9] Major Neil M. Lathrop was assigned as the Edwards Air Force Base project officer.

For each flight, Captain Stener taxied the B-50D into position on the dry lakebed. The T-33 was towed into position and attached to the towline. The Captain Stener slowly advanced the throttles of the four R-4360 engines. The B-50D inched forward, taking up the slack in the towline. Captain Stener pushed the throttles all the way forward to the stops, and the pair accelerated across the hard packed clay.

The B-50D lifted off the ground first. Once the B-50D lifted off the ground Captain Stener leveled off, retracted its landing gear, and accelerated to the take off speed of the T-33. With the T-33 trailing behind it, the B-50D climbed away from Rogers Dry Lakebed. If any unsatisfactory control or stability tendencies revealed themselves immediately after take-off, the airplanes could land straight ahead on the lakebed. Captain Newman in the T-33 concentrated on staying out of the prop-wash of the bomber as they performed the towing evaluation.

*Cranes hoist the Thunderjets to the wing tips of the Superfortress for ground tests of the redesigned coupling mechanism. (Bud Anderson)*

EF-84D 48-0641 hangs from a crane during ground tests of the coupling mechanism. (Bud Anderson)

EF-84D 48-0661 is coupled to the right wing tip of the EB-29A during ground tests of the redesigned mechanism. (Bud Anderson)

*The wing tip of the F-84 is locked to the B-29 in pitch and yaw. It is free to roll around the coupling. (Bud Anderson)*

*The lance on the wing tip of the F-84 is seated in the extended socket of the B-29 during ground tests. (Bud Anderson)*

*The pitch lock latch and the electrical and cockpit pressurization gas ports are housed in an aerodynamic fairing. (Bud Anderson)*

*Ports for the transfer of electrical power and pressurization gas are located between the lance and the pitch lock. (Bud Anderson)*

*The redesigned wing tip coupling mechanism reduced turbulence between the wing tips. (Bud Anderson)*

*The pitch lock latch fits into a notch in the wing tip of the Thunderjet. It engages the latch lock pin from the rear instead of the front. (Bud Anderson)*

*The variable pitch lock allowed the F-84 pilots to trim their fighters for wings level flight in cruise configuration.(Bud Anderson)*

The tests were concluded after eleven flights on March 8, 1951. Towed flight presented no particular problem to the pilot of the T-33, although it required his continuous attention to keep the airplane flying smoothly on the towline. Towed flight time totaled eighteen hours and thirty minutes over the course of the project.

#### Further development

The EB-29A and the two EF-84Ds remained at Wright Field following the contracted portion of the project. Wright Field pilots continued to evaluate additional aspects of wingtip-coupled flight.

Lights were installed on the wingtips of the Thunderjets to illuminate the receiver, and night couplings were accomplished. According to Bud Anderson, "The way that worked, a beam of light shined aft from the receiver and when you got the probe lined up directly behind for proper engagement the beam of light illuminated the probe and you knew you were in the right position for engagement. We learned this from the C47/Q14 night engagements."

### 1951 Flight Test

#### April 17, 1951, Transfer to Air Research and Development Command

Authority over the test aircraft was transferred from the Air Materiel Command to the Air Research and Development Command on April 17, 1951. The Air Research and Development Command assigned them to the Wright Air Development Center on June 22, 1951.

#### Side Entry

In early 1951, a new coupling mechanism was installed on the test aircraft to investigate a side entry method of coupling in flight. The coupling boom was extended to length of thirty-six inches for these tests.

As originally designed, the long boom turned out to be too flexible, and the receiver was too small. The project pilot was unable to achieve contact. The coupling mechanism was redesigned for further testing.[iii]

#### Probe and Drogue

By the end of August 1951, a drogue receiver was installed on a wing tip of the EB-29A. A matching probe was installed on an EF-84D. The first trial gave favorable results, but revealed that some features of the system could be improved. After a short period of modification, additional tests of the new coupling mechanism were performed.[iv]

Both EF-84Ds were transferred to the Republic Company plant at Farmingdale, New York in mid-September 1951. Republic technicians installed a gyroscopic control system to reduce the workload on the pilot of the parasite in coupled flight.

The EB-29A followed the Thunderjets to Farmingdale on November 16, 1951. The remainder of the MX-1016 wing tip coupling tests were conducted at Farmingdale.

#### *Automatic Flight Control*

Republic was awarded a contract to develop an automatic flight control system that would maintain the fighter in the proper attitude during wing tip coupled flight.

#### *Hardware development*

The contract with Republic was amended to include additional system refinements. These primarily involved supporting the systems of the EF-84D through the wing tip coupling mechanism. The wing tip coupling mechanism itself was substantially redesigned.

The roller bearing on the wing of the EF-84D was deleted from the design. The fairing into which the bearing seated on the wing of the EB-29A was also deleted.

A more substantial pitch lock latch was installed on the EB-29A. The revised latch retracted into an aerodynamic fairing on the upper surface of the wing tip. The latch rotated forward out of the wing of the EB-29A and seated on the rear end of the pitch lock latch pin, the opposite of the action of the original latch.

The wing tips of the Thunderjets were notched to receive the latch, eliminating the spring-loaded door of the earlier design.

The engines of the EF-84Ds were shut down in coupled flight. The electrical system of the EF-84D depended on the jet engine for its power. An alternative source of electricity was provided through the coupling. An interphone connection was also routed through the coupling.

Ports were installed on the wing tips of the EF-84Ds between the lance and the pitch lock to receive electrical and oxygen couplings from the wing of the EB-29A. Retractable mechanisms were installed in the wing tips of the EB-29A to mate electrical and oxygen connectors with the receivers in the wing tips of the EF-84Ds.The connectors accommodated the rotation of the EF-84Ds around the couplings.

Republic was directed to develop a system to transfer fuel from the EB-29A to the EF-84D through the coupling to replace the fuel used to climb to the rendezvous with the carrier. High-pressure gas to pressurize the cockpit of the EF-84D was to be provided through the coupling.

The complexity of the proposed system was daunting. Fuel, high-pressure gas, electricity, and a communications link all had to be routed through a mechanism that allowed the EF-84D freedom to roll about the wing tip of the EB-29A. The mechanisms had to connect automatically when the parasite coupled with the carrier. They had to disconnect instantly when the parasite disengaged.

A vacuum flask for the storage of liquid oxygen was to be installed in each EF-84D. As the liquid oxygen boiled off, it would provide a fifteen-hour supply of oxygen for the pilot.[v]

### *Later Operations, March - April 1953*

#### *Modifications*

Republic subcontracted the development of an autopilot to stabilize the EF-84D while it was coupled to the wing tip of the EB-29A. The autopilot sensed the angle of the connection between the airplanes and used the elevator of the EF-84D to control its flapping angle.

The wing tip seals of the modified aircraft were redesigned to reduce the turbulence that they induced.

The pitch locks were modified to allow the Thunderjet a limited range of motion in pitch. This relieved the twisting motion applied to the wing of the EB-29A by the application of the elevator of the fighter.

Doors were installed in the jet intakes of the EF-84D Thunderjets to shut down the airflow through the engine during coupled flight.

Data recording instruments were installed only in 48-661, the right-hand EF-84D. The left-hand EF-84D was not specially instrumented. This dictated that Major Anderson would conduct the initial evaluation of each operation while flying in the right-hand EF-84D.

#### *Resumption of flight tests*

Tip-tow flights resumed at the Republic plant in

*Full scale tests of a Boeing B-29 Superfortress and a pair of Republic F-84D Thunderjets began in 1950. (Bud Anderson)*

Farmingdale, New York in March 1953.

The primary pilot, Lt. Colonel Bud Anderson was no longer assigned to the Wright Air Development Center. His primary assignment was at the Pentagon. Due to his wing tip coupling experience, he was loaned back to the program.

The primary objective was to evaluate the newly developed automatic pilot for the Thunderjets. The function of the system was to roll the fighter toward a flapping angle of zero and to damp out any up and down oscillation of the fighter. The tests sought to determine the proper settings to damp the oscillations as quickly as possible.

The autopilot was tuned through trial and error. Once the pilot engaged the autopilot, he would momentarily deflect the elevator to displace the fighter from wings level, then release the stick, and observe how long it took for the system to damp out the resulting oscillation. The damping settings would be adjusted and the pilot would make another control pulse. This would be repeated until the correct settings were determined.

Due to time and money constraints, there were no instruments to indicate the flapping angle of the wing or the elevator deflection that the autopilot was going to command when it was engaged.

On the first flight of the resumed tests, Anderson found that his EF-84D was not drawing electrical power from the bomber. Major Davis found that the electrical connection to his plane was working properly, but it did not have the necessary data recording instruments to evaluate the autopilot.

Anderson returned to his primary assignment at the Pentagon while the technicians at Republic worked on the electrical power problem in his Thunderjet. When they were satisfied that the problem was fixed, Anderson flew back to Farmingdale. When Anderson coupled his EF-84D to the wing of the EB-29A, it failed to draw power again.

*Investigators hoped to evaluate cruise flight with the jet engines of the F-84s shut down and inlet doors closed. This required the B-29 to provide electrical power to the fighter. (Bud Anderson)*

Anderson found himself shuttling back and forth between his Pentagon job and the test flights at Republic. Each time he returned to Farmingdale and coupled his EF-84D to the wing tip of the bomber, his fighter failed to draw electrical power. Major Davis' fighter was functioning properly but lacked instrumentation. Back at the Pentagon, Anderson's boss expressed his concern about his time away. His boss would have been far happier if flight tests were performed on schedule, with no delays due to unexpected problems.[vi]

To evaluate the repairs to the electrical connection in Anderson's Thunderjet, technicians at Republic hoisted it up to the wing tip of the Superfortress with a crane.

They inserted the lance into the receiver and retracted it into place. The power indicator light came on in the cockpit. They rolled the EF-84D around the wing tip and the electrical power stayed on.

### *April 24, 1953 Final Mission*

The sixth attempt to test the autopilot was flown on April 24, 1953. The EB-29A was piloted by Captain J. J. Vapenick, and Captain H. C. McClanahan. The crew included Master Sergeant D. D. Shaffer, Master Sergeant Claude Chemp, and Technical Sergeant J. R. Clarkson on detached service from the Wright Air Development Center.

The crews were determined to test the autopilot. Before the flight, the pilots had agreed that Anderson would couple his fighter to the bomber for the first test. If he failed to get an electrical connection, he would disengage from the bomber, and Davis would link up to make the test.

The test planes took off at 4:17 in the afternoon.

The EB-29A held a steady course at 25,000 feet altitude over Peconic Bay. Anderson lined up with the wing tip and locked his probe into the receptacle. The

*Full scale tests of a Boeing B-29 Superfortress and a pair of Republic F-84D Thunderjets began in 1950. (Bud Anderson)*

Farmingdale, New York in March 1953.

The primary pilot, Lt. Colonel Bud Anderson was no longer assigned to the Wright Air Development Center. His primary assignment was at the Pentagon. Due to his wing tip coupling experience, he was loaned back to the program.

The primary objective was to evaluate the newly developed automatic pilot for the Thunderjets. The function of the system was to roll the fighter toward a flapping angle of zero and to damp out any up and down oscillation of the fighter. The tests sought to determine the proper settings to damp the oscillations as quickly as possible.

The autopilot was tuned through trial and error. Once the pilot engaged the autopilot, he would momentarily deflect the elevator to displace the fighter from wings level, then release the stick, and observe how long it took for the system to damp out the resulting oscillation. The damping settings would be adjusted and the pilot would make another control pulse. This would be repeated until the correct settings were determined.

Due to time and money constraints, there were no instruments to indicate the flapping angle of the wing or the elevator deflection that the autopilot was going to command when it was engaged.

On the first flight of the resumed tests, Anderson found that his EF-84D was not drawing electrical power from the bomber. Major Davis found that the electrical connection to his plane was working properly, but it did not have the necessary data recording instruments to evaluate the autopilot.

Anderson returned to his primary assignment at the Pentagon while the technicians at Republic worked on the electrical power problem in his Thunderjet. When they were satisfied that the problem was fixed, Anderson flew back to Farmingdale. When Anderson coupled his EF-84D to the wing of the EB-29A, it failed to draw power again.

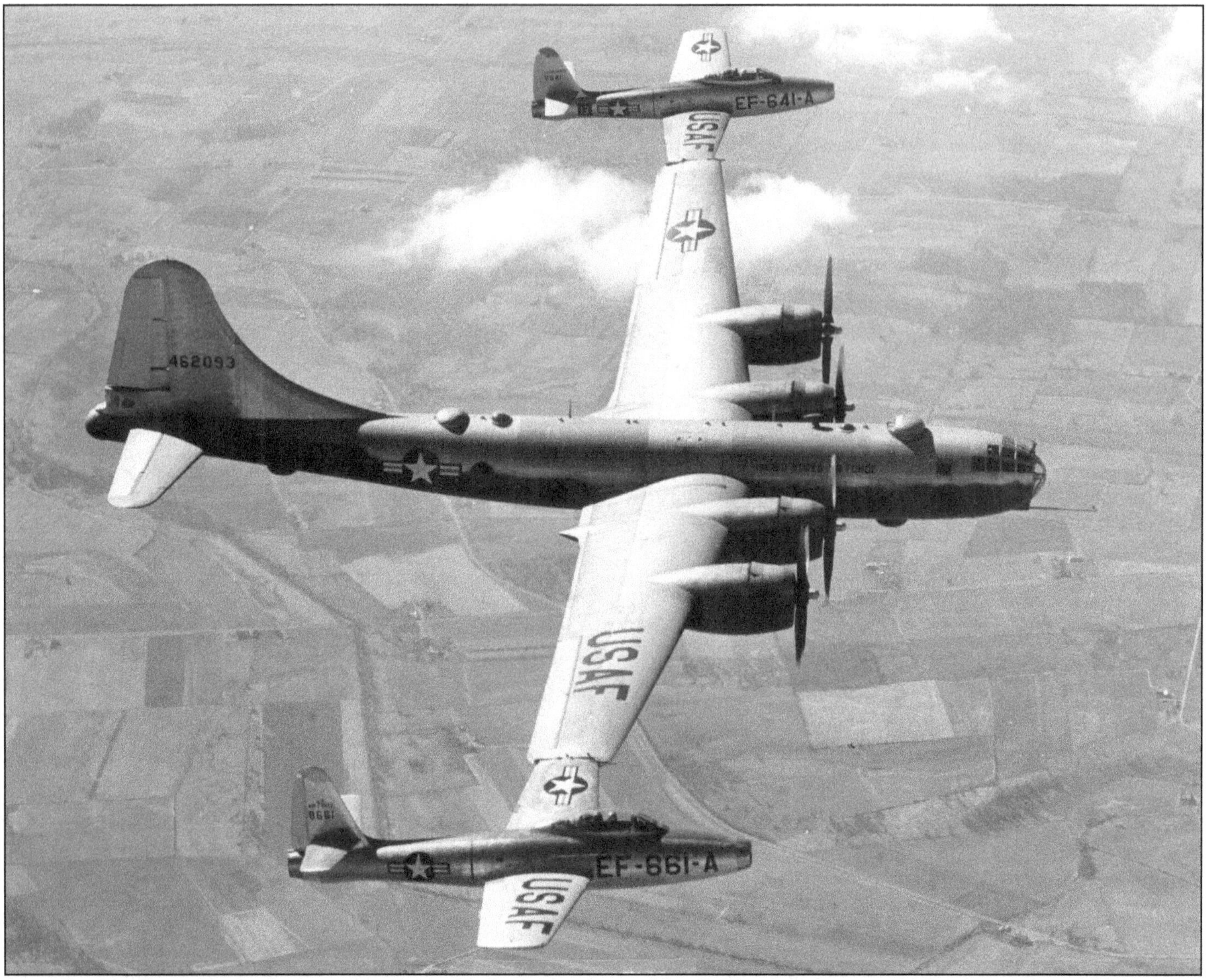


*Investigators hoped to evaluate cruise flight with the jet engines of the F-84s shut down and inlet doors closed. This required the B-29 to provide electrical power to the fighter. (Bud Anderson)*

Anderson found himself shuttling back and forth between his Pentagon job and the test flights at Republic. Each time he returned to Farmingdale and coupled his EF-84D to the wing tip of the bomber, his fighter failed to draw electrical power. Major Davis' fighter was functioning properly but lacked instrumentation. Back at the Pentagon, Anderson's boss expressed his concern about his time away. His boss would have been far happier if flight tests were performed on schedule, with no delays due to unexpected problems.[vi]

To evaluate the repairs to the electrical connection in Anderson's Thunderjet, technicians at Republic hoisted it up to the wing tip of the Superfortress with a crane.

They inserted the lance into the receiver and retracted it into place. The power indicator light came on in the cockpit. They rolled the EF-84D around the wing tip and the electrical power stayed on.

***April 24, 1953 Final Mission***

The sixth attempt to test the autopilot was flown on April 24, 1953. The EB-29A was piloted by Captain J. J. Vapenick, and Captain H. C. McClanahan. The crew included Master Sergeant D. D. Shaffer, Master Sergeant Claude Chemp, and Technical Sergeant J. R. Clarkson on detached service from the Wright Air Development Center.

The crews were determined to test the autopilot. Before the flight, the pilots had agreed that Anderson would couple his fighter to the bomber for the first test. If he failed to get an electrical connection, he would disengage from the bomber, and Davis would link up to make the test.

The test planes took off at 4:17 in the afternoon.

The EB-29A held a steady course at 25,000 feet altitude over Peconic Bay. Anderson lined up with the wing tip and locked his probe into the receptacle. The light that indicated the status of the wing tip electrical connection did not illuminate. He gave up the attempt to test the autopilot with his Thunderjet and disconnected from the wing tip of the Superfortress.

At approximately 5:45 PM, Major John Davis brought the left-hand EF-84D into position and linked wingtips with the EB-29A. He transmitted, "Autopilot coming on" and then switched the autopilot on. It commanded hard nose-up elevator. Davis' Thunderjet rolled rapidly to the right onto the outer wing panel of the Superfortress. As the fighter rotated about the wing tip of the bomber, the explosive bolts fired to jettison the Thunderjet, but not in time to prevent the collision between the planes.

The left wing of the EB-29A failed outboard of the engines. The nose of the EF-84D broke away from the rest of the airplane.

The pieces of the two airplanes tumbled downward, streaming flames and smoke. To get out of the rear compartment of the Superfortress, the crew has to go through the aft bulkhead into the tail and then bail out the exit hatch. Only Master Sergeant Chemp in the EB-29A managed to work his way to an emergency exit and bail out. There was not enough time for his parachute to open before he and the bomber hit the water of Great Peconic Bay about 3-1/2 miles south of Robins Island. The bomber sank with four crewmembers still on board, leaving a large oil slick on the surface.

The aft section of the EF-84D landed in a dense forest on the south shore of the bay. Its fuel tanks ruptured, and their contents erupted in a ball of flame.

Major Anderson watched the tragedy unfold from his Thunderjet on the right side of the Superfortress. If his airplane had not suffered its persistent electrical malfunction, he likely would have been the pilot to discover the autopilot's fatal flaw.

*The wing span of the three airplanes totaled 213 feet. (Bud Anderson)*

On the ground, Nick Nowaski and Frank Catricone, operators at the McKay Marin Receiving Station outside Southampton heard the airplanes fly overhead. They did not witness the crash, but they saw the smoke rising from the woods where the Thunderjet came down, about one mile from their shack.

Alick Gerard, the grounds keeper at the National Golf Links of America witnessed the crash.

Volunteer firemen from the town of North Sea drove to the receiving station, where Nowaski and Catricone directed them toward the post-crash fire. They found the Thunderjet completely destroyed by impact and fire.

Dr. John Nugent, the local coroner, was transported to the site of the Thunderjet crash in an Army truck. He found Davis' body a short distance from the charred wreckage. His body was removed to David De Friest's morgue in Southhold.[vii]

Lieutenant Thomas Ward, a Civil Air Patrol pilot was flying a privately owned amphibious airplane in the area at the time of the crash. He landed his floatplane in the middle of the oil slick, but found no trace of the crewmen.

Air Force rescue planes arrived at the crash site after sunset.

The crash boat from the Suffolk County Air Force Base at West Hampton Beach and three Coast Guard vessels from Shinnecock Station at Hampton Bays arrived about 4:00 in the morning after the crash. The vessels searched through the night for survivors in vain. The Air Force crash boat was disabled but the three Coast Guard vessels continued the search into the next day.

Master Sergeant Chemp's body was recovered at 2:35 that afternoon. He was identified initially by the name stamped on his parachute pack.

The vessels secured their search at 4;00 PM and resumed searching first thing the next morning.

Major Lester L. Applegate of Suffolk County Air Force Base was put in charge of salvage operations. Divers and salvage equipment were brought in from Bayonne, New Jersey to raise the hulk of the EB-29A.

No mention of the wing tip coupling experiment was made in the New York Times account of the crash.

It was reported that a factor contributing to the crash was the division of authority between the Wright Air Development Center, Republic Aviation, and Republic's autopilot subcontractor. None had explicit responsibility for integrating the safety efforts of the team.[viii]

Major Davis had been about to begin Project FICON flight tests of the swept wing YRF-84F Thunderstreak prototype. Lt. Colonel Anderson was drafted back into Project FICON. After making his initial evaluation of the YRF-84F in May 1953 he checked out Major Clyde Good and handed the program off to him.

EF-84D, 48-661 returned to the Wright Air Development Center in June 1953. It remained there for two more years. In July 1955 it was transferred to the Ogden Air Materiel Area at Hill Air Force Base, Utah. In February 1956 its designation was changed to JF-84D. It was dropped from the inventory of the Air Force in March 1957.

Wing tip coupling experiments resumed under the auspices of Project Tom-Tom at the Convair plant at Fort Worth in 1955 using Convair JRB-36F 49-2707, which had been the FICON test bed, and Republic RF-84F Thunderflashes 51-1848 and 51-1849.

### References


[i]Flight Tests of Project MX-1016; Republic Aviation Corporation, Engineering Flight Report 88, January 16, 1951

[ii]Towing time log, Fairchild Aviation Corporation, Report No EFR-88, Appendix C, January 16, 1951

[iii]Research and Development Information Report 51S-36820; Engineering Division, Air Materiel Command; February 28, 1951

[iv]Research and Development Information Report 51S-132084; Engineering Division, Air Materiel Command; August 30, 1951

[v]Research and Development Information Report 51S-18033; Engineering Division, Air Materiel Command; January 31, 1951

[vi]To Fly and Fight; Col. Clarence E. "Bud" Anderson with Joseph P. Hamelin; Pacifica Press; 1990; Page 215

[vii]Jet and B-29 Collide, 6 die at Southampton, New York Times, April 25, 1953.

[viii]To Fly and Fight; Col. Clarence E. "Bud" Anderson with Joseph P. Hamelin; Pacifica Press; 1990; Page 217

*Project Tom-Tom continued the evaluation of wing tip coupling using a Convair RB-36F and a pair of Republic RF-84F Thunderflashes from 1953 to September 1956. (General Dynamics photo 26-2290)*

## Section 3: Project Tom-Tom

### *Wing tip-Coupling Wind Tunnel Experiments*

A series of wind-tunnel investigations were conducted at the NACA Langley Aeronautical Laboratory in Virginia in support of the wing tip-coupling concept.

#### *Flight Tests of a Model having Self-Supporting Fuel-Carrying Panels Hinged to the Wing Tips, November 1949*

It would be necessary to conduct wind tunnel tests to verify the feasibility of Dr. Vogt's proposal to carry fuel tanks in airfoil shaped panels hinged to the wing tips of an airplane. Mathematical aerodynamic models of the time were not sufficiently sophisticated to adequately analyze the behavior of an airplane equipped with hinged wing tip panels.

A model that could be equipped with a variety of hinged wing tip panels was tested in the free-flight wind tunnel at the NACA Langley Aeronautical Laboratory in late 1949. The principle investigators were Robert E. Shanks and David C. Grana. Their objective was observe the flying qualities and evaluate the difficulty of controlling the complex articulated model.

The tests utilized a model airplane of normal configuration to which a variety of self-supporting panels could be attached at the wing tips. The wingspan of the basic model was 42½ inches. The aspect ratio of the wing was 5.06.

One pair of panels was designed to attach rigidly to the wing of the model. Other panels were equipped with hinges to allow them to float freely, supported by the lift that they generated. The panels were weighted to have approximately the same wing loading as the main wing. Each pair of panels weighed approximately one third as much as the airplane model itself.

The hinge line of each pair of panels was set at a different angle relative to the chord of the wing. One pair had the hinge line parallel to the wing chord. The hinge line of other pairs skewed outward at the front edge by 20 degrees and 45 degrees.

Skewing the hinge line of the wing tip panels provided a restoring force when the panel rolled away from the wings-level position. When the panel rolled upward, its leading edge pitched downward. The reduction in the lift of the panel pushed it back down. When the panel rolled downward, its leading edge pitched upward, increasing its lift and pushing it back up.

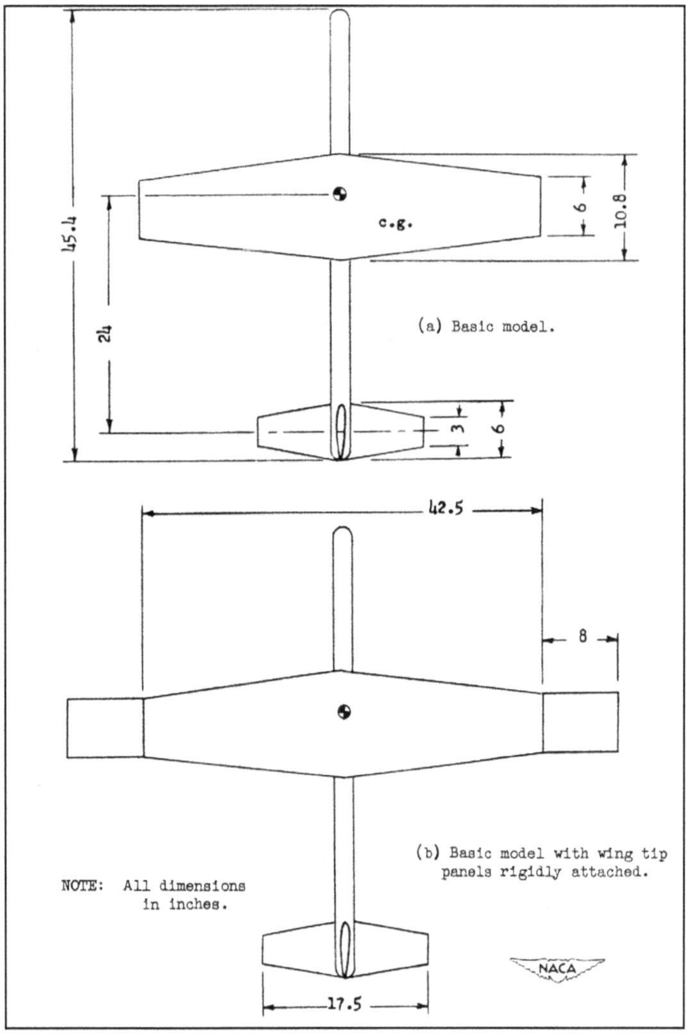


*Sketch of the model used in the tests with and without wing tip panels. (NACA RM L9I07a)*

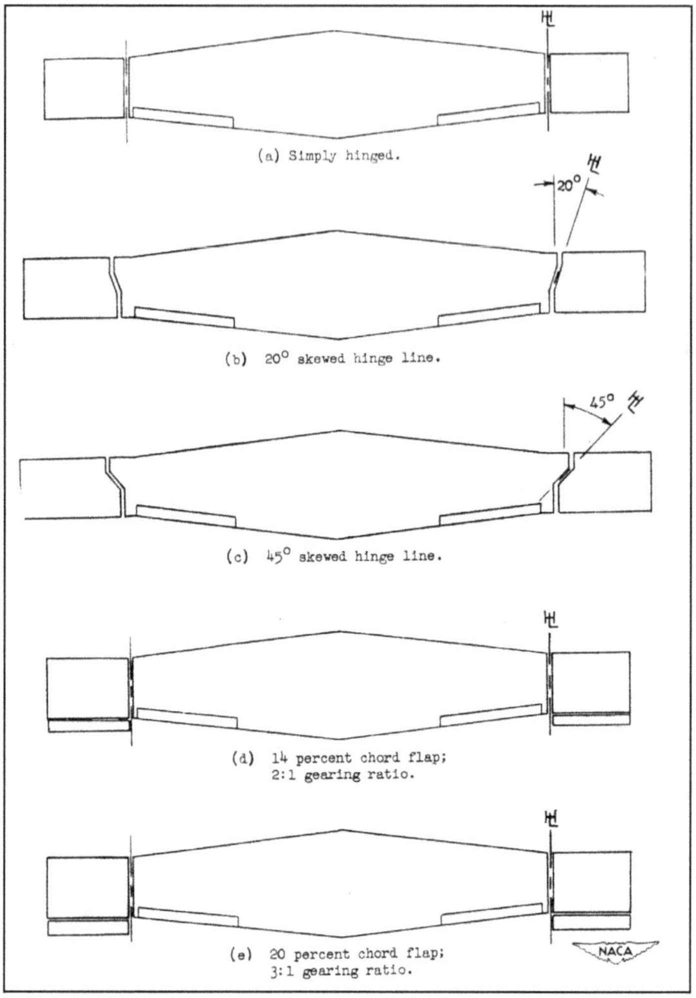


*Sketch of the hinged-wing tip-panel configurations tested. (NACA RM L9I07a)*

The investigators also tested wing tip panels equipped with full span trailing edge flaps. A mechanical linkage deflected the flaps as the panel rotated relative to the wing tip. The deflection of the flap acted to return the panel to the wings level position. The ratio of flap deflection to the angle of rotation of the panel could be adjusted between test runs.

Flaps of two different sizes were tested. One set of flaps was 14% of the size of the panel and the other was 20% as large.

The investigators did not rigorously investigate the performance benefit derived from the increased aspect ratio. The relatively high pressure under the wing forced air to flow through the gap between the wing tip of the model airplane and the free-floating panels, reducing the aerodynamic efficiency of the composite model. The full benefit of the increased aspect ratio would only be realized if the gap was sealed. The investigators performed tests of the wing tip panels that had the hinges parallel to the wing chord with the gap at the wing tip of the model both sealed and unsealed to make a rough evaluation of the effect of sealing the gap.

The first test was performed without any wing tip extensions to make baseline stability and controllability evaluations of the model airplane by itself. Then wing tip panels were attached rigidly to the wing tips of the airplane model. As would be expected, the rigid panels caused the model to respond more slowly to the application of roll control inputs. The model with the rigidly attached panels also displayed a tendency to oscillate left and right slightly. The rigid panels made the model slightly less stable because they increased the moments of inertia of the model in roll and yaw, but the size of the stabilizing surfaces at the tail had not been increased to compensate for the decreased stability.

The investigators then installed free-floating wing panels with hinges aligned parallel to the chord of the wing. The hinged wing tip panels added significantly to the complexity of the behavior of the model. Any disturbance of the position of the free-floating panels started a flapping motion that continued for several oscillations before dying out. Each of wing tip panels weighed one sixth as much as the model airplane, so they pulled the model's wing tips up and down as they flapped. The panels flapped at a higher frequency than the airplane's natural rolling frequency, so at some times they reinforced the rolling of the model and at other times they against the rolling of the model. The pilot of the model was able to control its motions in a general fashion, but it persistently jerked back and forth in a chaotic fashion.

Sealing the gaps between the wing tips of the model and the hinged panels did not improve the behavior of the model except to reduce its drag somewhat.

Panels were attached to the wing tips of the model by hinges that were skewed relative to the chord of the wing to investigate whether the skewed hinge would reduce the objectionable flapping motion of the panels. Hinges skewed at 20 degrees and 45 degrees accomplished a degree of improvement, but the behavior of the composite model was still considered unsatisfactory.

The results obtained by installing full span flaps on the wing tip panels were much better than those obtained with the skewed hinge. The flaps could be adjusted to move one, two, or three degrees for every degree that the panel rolled relative to the wing tip. At the lower ratios, the flaps produced better behavior than the hinge skewed at 45 degrees, although the general behavior of the model was still not satisfactory. When the larger flaps were adjusted to move three degrees per degree of roll the wing tip panels followed the motions of the airplane model quite closely. The jerky rolling motions were nearly eliminated.

The tests indicated that it would be feasible to increase the range of a bomber by attaching fuel-filled panels to its wing tips with hinges. Sufficient stabilization should be obtainable using full span flaps mechanically linked to the rolling movement of the panels relative to the wing tips.[i]

#### *B-50 Floating Wing Tip*

In early 1950, engineering calculations were performed to determine the forces that unmanned floating wing tip panels would impose on a Boeing B-50 Superfortress. The greatest forces would be imposed on the wing tips of the Superfortress during emergency landings, when the floating panels would be full of fuel.

The system under consideration was designed with a two-point hinge which allowed the floating panel to roll about the wing tip, but did not allow the panel to rotate about its pitch or yaw axes. Fixed to the bomber's wing tip in this way, the floating panels applied excessive twisting forces through the coupling, particularly in the event of an emergency landing.

Consideration was given to switching to a single-point attachment. Allowing the floating panels to pitch and yaw would eliminate the high torque loading on the wing of the bomber. Unfortunately, adopting the single-point hinge would require a complete revision of the dynamic analysis of the system.[ii]

#### *B-29/F-84 Wing Tip Coupled Wind Tunnel Tests, January 1951*

In late 1950, the Engineering Division of the Air Materiel Command conducted tests of a scale model of the B-29 with a pair of wing tip coupled F-84s in their 20-foot wind tunnel for comparison with the results obtained during the full-scale, wing tip coupling flight tests.[iii]

#### *Free-Flight-Tunnel Investigation of the Dynamic Lateral Stability and Control Characteristics of a tip-to-tip Bomber-Fighter Coupled Airplane Configuration, April 1951*

Investigators at the NACA Langley Aeronautical Laboratory conducted wind tunnel tests to evaluate a method to automatically control the attitude of a small airplane coupled to the wing tip of a larger airplane in late 1950. The principle investigators were Charles V. Bennett and Robert B. Cadman.

A very simplified model of a large airplane with a smaller airplane attached to each wing tip was flown in the Langley free-flight wind tunnel. The large airplane represented a bomber and the smaller airplane models represented parasite fighter airplanes. The primary objective of the investigation was to determine the effectiveness of a mechanical linkage that moved the outboard aileron of the fighters to keep them aligned with the wing of the bomber.

The angle between the wing of the bomber and the wing of the fighter was referred to as the flapping angle. When the wings of the fighter were lined up with the wings of the bomber, the flapping angle was zero.

A mechanical linkage was installed to convert the flapping angle of the fighter to a displacement of its outboard aileron that would push the fighter back toward a flapping angle of zero. The linkage could be adjusted to vary the amount of aileron deflection produced by a given flapping angle. At the lowest adjustment, the angle of deflection of the aileron was equal to the flapping angle of the fighter. At the highest ratio, the deflection of the aileron was nine times as great as the flapping angle. During the full-scale tests conducted by the WADC with the EB-29B Superfortress and F-84D Thunderjets in 1950 the aileron was geared to move only 0.6 times the amount of the flapping angle.

Each of the fighter models was attached to a wing tip of the bomber model through a series of three pivots. The pivot that was attached directly to the bomber's wing tip allowed the fighter to rotate in yaw. The middle pivot allowed the fighter to roll around the wing tip, and the outboard pivot allowed fighter to rotate in pitch. Each of the pivots could be locked in position independently of the others.

During the initial tests, the smaller airplanes were fixed rigidly to the wing tips of the larger airplane model. When the fighters were fixed to the wing tips of the bomber, the composite model behaved like a conventional airplane. The roll response was slower and its lateral stability was reduced when compared to the bomber model alone.

In subsequent tests, the fighters were given progressively more degrees of freedom of rotation. First, the fighters were allowed the freedom to rotate in

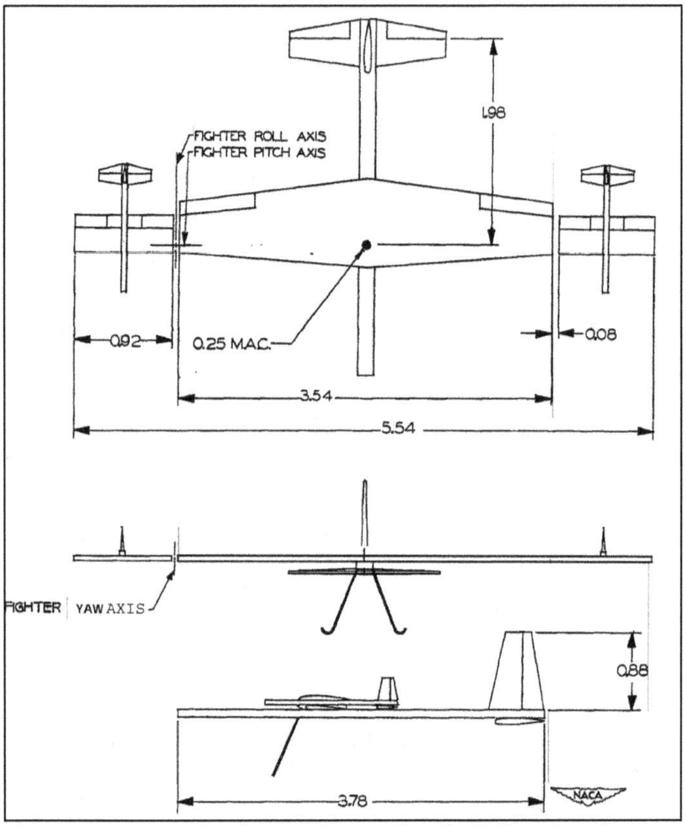


*Three-view sketch of wing-tip to wing-tip fighter-bomber configuration tested in the Langley free-flight tunnel. All dimensions are in feet. (NACA RM151a12)*

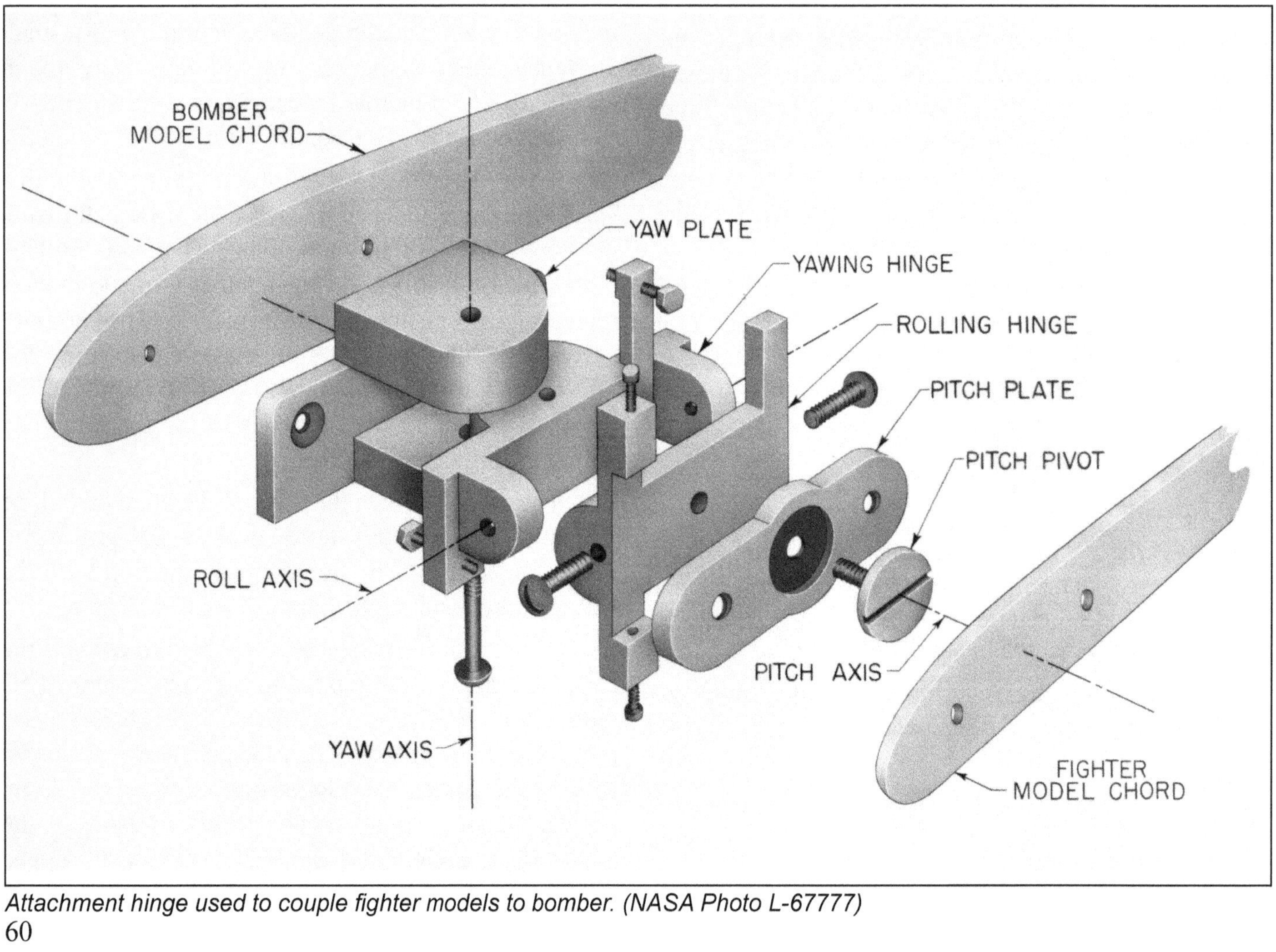


*Attachment hinge used to couple fighter models to bomber. (NASA Photo L-67777)*

roll. Then they were also allowed to rotate in pitch as well as roll. In the course of the final tests, the fighters were free to rotate about all three axes.

In each sequence of tests, the gearing ratio of the outboard aileron control was set first to move the aileron through an angle equal to the flapping angle of the smaller model. Between tests, the gearing ratio was then increased in steps until the flap moved through an angle eight times as great as the flapping angle.

During the first series of tests, the smaller models were fixed in pitch and yaw and only allowed to roll around the coupling mechanism. This corresponded to the general configuration tested with the EB-29B Superfortress and F-84D Thunderjets. A gearing ratio of one to one, which was actually greater than that used in the full-scale tests, provided only minimal damping of the flapping motions of the smaller models. The fighters wobbled nearly randomly and made it nearly impossible for the pilot controlling the larger model to maintain smooth, wings-level flight. The behavior of the model was only satisfactory at the maximum gearing ratio.

An additional test was run at the maximum gearing ratio with a ½-inch extension added to the smaller model's aileron. The extra area of the aileron produced further improvement in the behavior of the models. The pilot made abrupt control inputs to demonstrate that the smaller models closely followed the bomber through banking maneuvers. The behavior of the models at the maximum gearing ratio with the aileron extension was considered to be nearly as good as when they were fixed rigidly to the bomber's wing tips.

The pitch axis of each of the coupling mechanisms was unlocked for the next series of tests. The smaller models were then free to roll and pitch around the wing tip of the bomber. As the smaller models pitched up and down, the lift generated by their wings constantly varied. This additional variation in the lift of the smaller models induced even more flapping than when they were fixed in pitch.

Since the smaller models were moving independently of each other, the variation in their lift produced unpredictable variable asymmetric forces on the bomber. As the gearing ratio was increased, the flapping angles of the smaller models decreased, but the flight behavior of the bomber did not improve as much as it had at the same gearing ratio during the previous test. Even so, the pilot felt that the behavior of the models was satisfactory at the maximum gearing ratio with the aileron extension.

All three axes of the coupling mechanisms were freed for the last series of tests. This corresponded to the configuration tested by the WADC using the C-47 and PQ-14 in 1949. The additional freedom of rotation produced even greater flapping by the smaller models. Rotation about any one axis produced forces that affected the rotation of the smaller models about the other two axes. The resulting movements were chaotic and difficult to correct for at the lower gearing ratios. As in the previous test, the behavior of the composite model improved as the gearing ratio was increased. At the maximum gearing ratio with the aileron extension, the pilot felt that the flight behavior of the models was satisfactory.

The test observers concluded that the flight behavior of the models could be made satisfactory by applying a sufficiently high gearing ratio. This was true for models attached with one, two, or three degrees of freedom. The gearing ratio necessary to properly damp out the flapping motion of the fighters was much higher than the ratio that had been used during the full-scale tests.

Although the general flight behavior of the composite models was satisfactory even when the flapping of the smaller models was lightly damped, the test observers concluded that higher gearing ratios would be necessary to provide a comfortable ride for the pilot riding in the smaller airplane. Although flight tests with full-scale hardware had demonstrated that the pilot of the smaller airplane could control its flapping, it was not realistic to expect the pilot to continually control the airplane for the duration of an actual mission.[iii]

### *Free-Flight-Tunnel Investigation of the Dynamic Lateral Stability and Control Characteristics of a High-Aspect-Ratio Bomber Model with Self-Supporting Free-Floating Fuel Tanks Attached, August 1951*

The initial wind tunnel tests of wing tip coupled fuel tanks that were conducted at NACA Langley Aeronautical Laboratory had utilized a very simplified, almost schematic, composite airplane model. Investigators desired to conduct tests of a composite airplane model that conformed more closely to the configuration of a full-scale composite airplane system.

Charles Bennett and Robert Cadman conducted a series of tests of a bomber model with high aspect ratio wings that had an airfoil-shaped panel attached to each of its wing tips in the Langley free-flight wind tunnel in 1952. Each wing tip panel supported a pod that

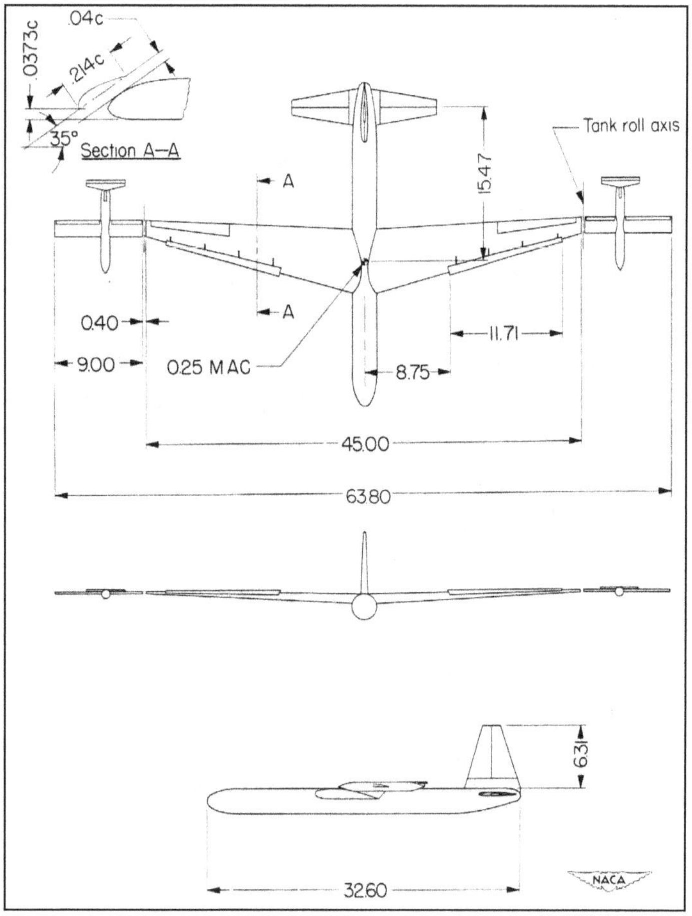

*Three-view sketch of the high-aspect-ratio bomber model with fighter models attached. All dimensions are in inches. (NACA RM151e17)*

represented a fuel tank. The wing loading of the fuel tanks was approximately the same as the wing loading of the bomber model.

An interesting feature of the wing tip tanks was that they were asymmetric. The inboard lifting surface had a shorter span than the outboard surface. This was because the lift distribution across the lifting surface was affected by the proximity of the bomber model's wing tip. The lifting surface that was closer to the bomber's wing tip produced more lift per unit area.

The lifting surfaces of the fuel tanks were equipped with full span trailing edge flaps. As in the previous experiments, a mechanical linkage deflected the flaps by an amount proportional to the flapping angle of the tank to stabilize the tank relative to the wing of the bomber model. The amount of flap deflection could be set to 3.33 times or 4.75 times the flapping angle.

The basic flaps were 20% of the size of the tank's lifting surface. Flap extensions could be installed between test runs to increase the flap area to 31% or 42% of the lifting surface area.

The bomber model had a wingspan of 45 inches and an aspect ratio of 10.76. Initial tests were made of the bomber alone to collect baseline stability and controllability data. The wing tip tanks were installed rigidly for some tests and free to roll around the wing tip of the bomber model in others.

The effect of the rigidly attached wing tip tanks was similar to that for the rigid wing tip panels in earlier tests. The response to aileron control inputs was slower and the stability of the composite airplane was somewhat reduced.

While the flaps were geared to move 3.33 times the flapping angle, the flapping motions of the wing tip tanks were only lightly damped, regardless of the size of the stabilizing flaps. The flight behavior of the model with was unsatisfactory. When the controls of the bomber were abruptly displaced for a moment to

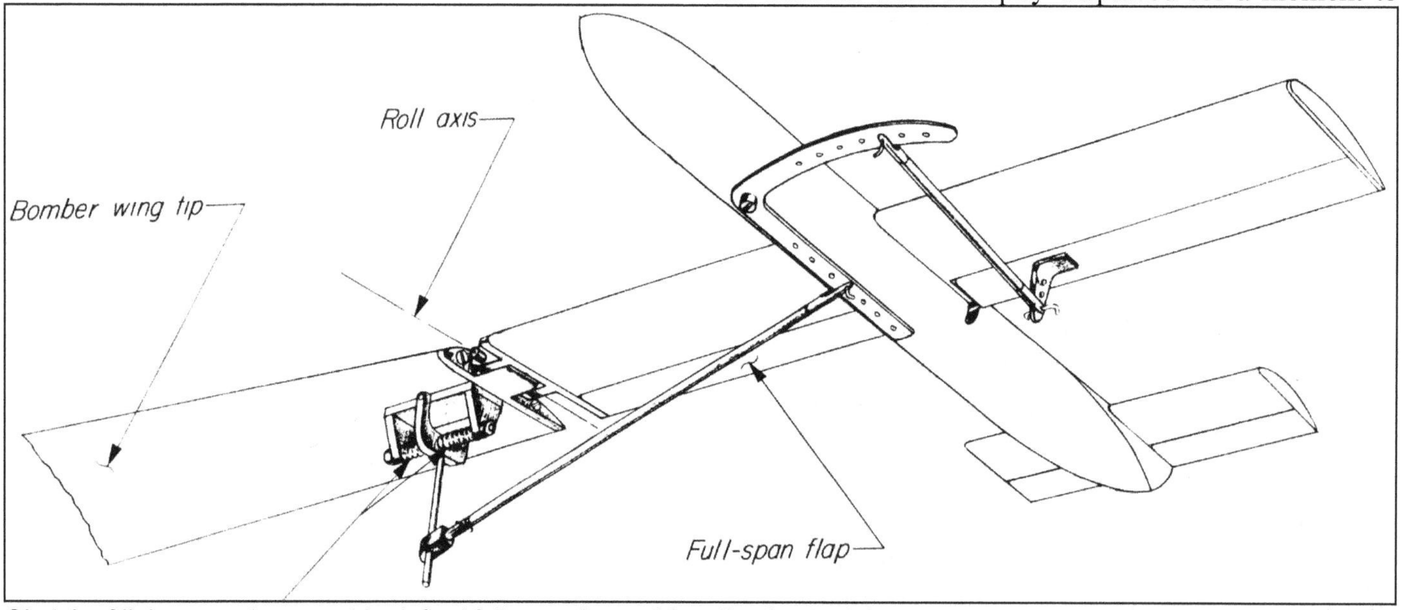

*Sketch of linkage system used to deflect full-span flaps of free-floating tank in response to the relative angle of bank of the tank with respect to the bomber. (NACA RM151e17)*

simulate gust loads, the tanks would begin flapping. The flapping of the tanks applied variable asymmetric loads to the bomber that caused it to wobble. The wobbling of the bomber then drove the tanks to continue flapping. The pilot controlling the bomber model found that the flapping of the tanks alternately damped out and pumped up the wobble of the bomber in a chaotic fashion.

When the gear ratio of the flaps was adjusted to 4.75 the flapping angle, the pilot of the model felt that the objectionable wobbling was reduced compared to the lower gear ratio for each flap size. The behavior of the composite model was considered satisfactory only when the largest flap extensions were installed.[iv]

#### ***Free-Flight-Tunnel Investigation of the Dynamic Lateral Stability and Control Characteristics of a High-Aspect-Ratio Bomber Model with a Sweptback-Wing Fighter Model Attached to each Wing tip, September 1952***

Initial wind tunnel tests of fighters coupled to the wing tips of a bomber had utilized a simplified composite airplane model. Investigators at NACA Langley Aeronautical Research Laboratory desired to conduct tests of a composite airplane model that was more representative of actual Air Force hardware. The principle investigator was Charles Bennett. Peter C. Boisseau was co-author of the NACA research memorandum that described the tests.

Bennett and Boisseau conducted a series of tests of a bomber model with high aspect ratio wings that had a fighter model with swept wings attached to each of its wing tips in the Langley free-flight wind tunnel in 1952. This configuration corresponded generally to the configuration of the proposed B-36/RF-84F composite airplane to be evaluated in the Tom-Tom project. The tests used the same bomber model with high aspect ratio wings that had been used for the free-floating fuel tank tests the previous year.

The fighter models were free to roll around the wing tips of the bomber model. The angle of pitch of the fighter model relative to the bomber model could be adjusted between flights. Adjusting the angle of pitch changed the amount of lift generated by the wing of the model at any particular air speed. This allowed the investigators to trim the fighter model to fly at the desired flapping angle.

A mechanical linkage moved the outboard aileron of each of the smaller models in response to its rolling movement about the wing tip of the bomber model. The gearing ratio provided by the linkage could be varied from 1.7 to 5.4. In addition, the aileron could be

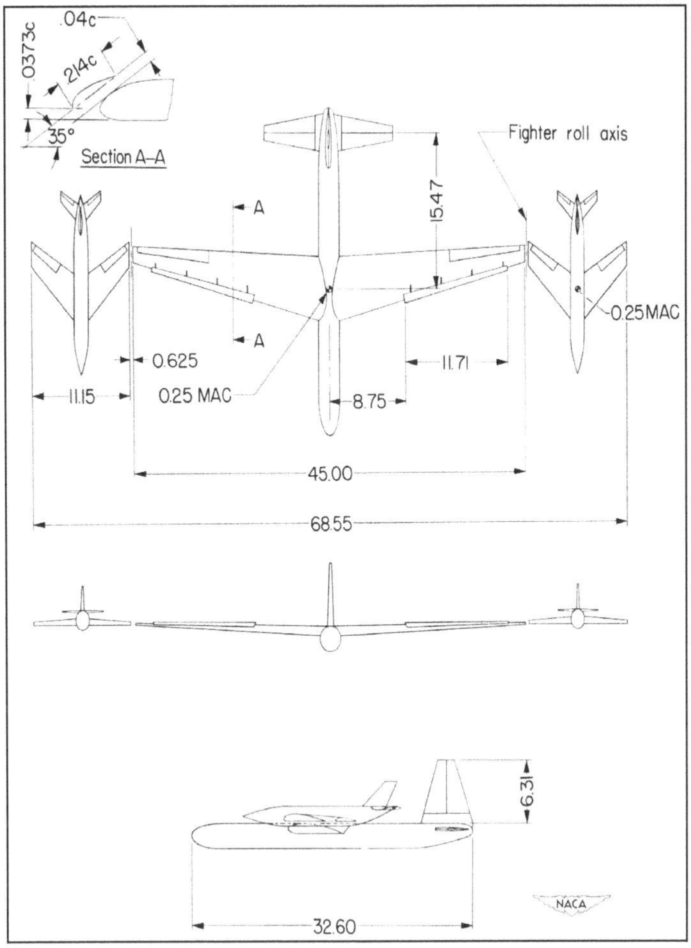


*Three-view sketch of the high-aspect-ratio bomber model with fighter models attached. All dimensions are in inches.(NACA RM152e08)*

fitted with an extension that added 38% to its area tom improve its control authority.

The position of the attachment point for each of the fighter models could be changed to adjust the position of the fighter model relative to the wing tip of the bomber model. The most obvious location to attach the fighter, where the wing tips of the two airplanes were even, placed the aerodynamic center of the fighter ahead of the wing of the bomber. When the fighter model was displaced in pitch, it twisted the wing of the bomber, further increasing the pitch of the fighter. The positive feedback to the pitch angle of the fighter had a deleterious effect on its stability. Mounting the fighter farther aft on the wing tip placed the fighter's aerodynamic center closer to the wing of the bomber. This reduced its leverage and the amount of bending of the bomber's wing.

As in the earlier free-flight wind tunnel tests, increasing the gearing ratio and area of the outboard aileron improved the damping of the fighter's flapping motions and improved the flying behavior of the composite airplane.

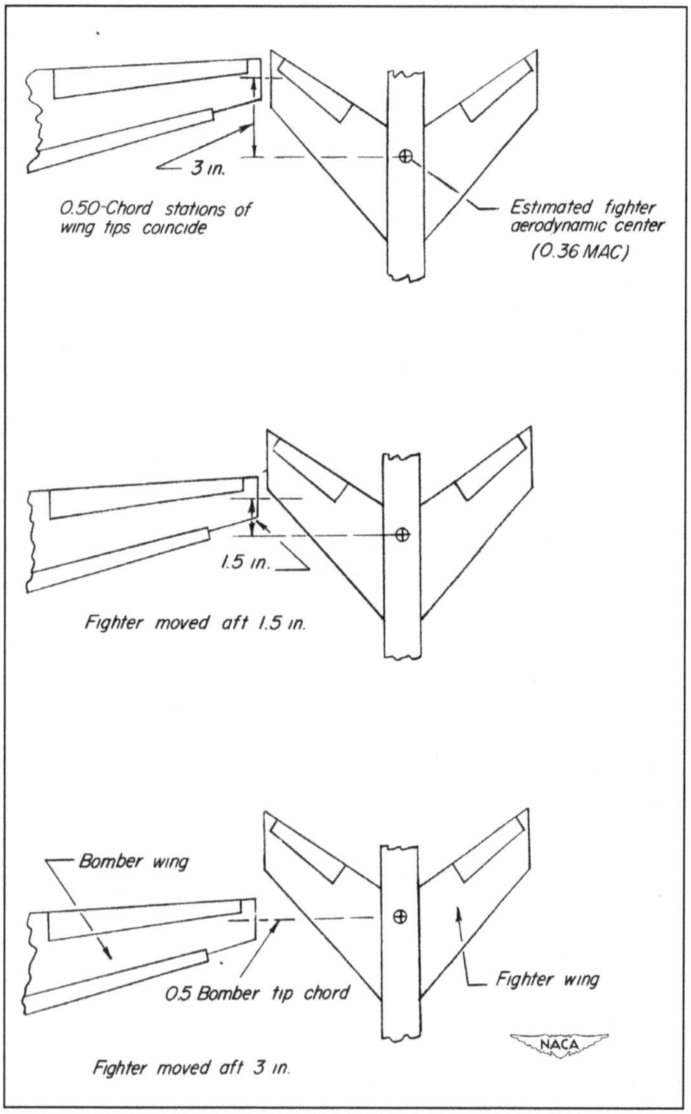


*Longitudinal positions of fighter models tested. (NACA RM152e08)*

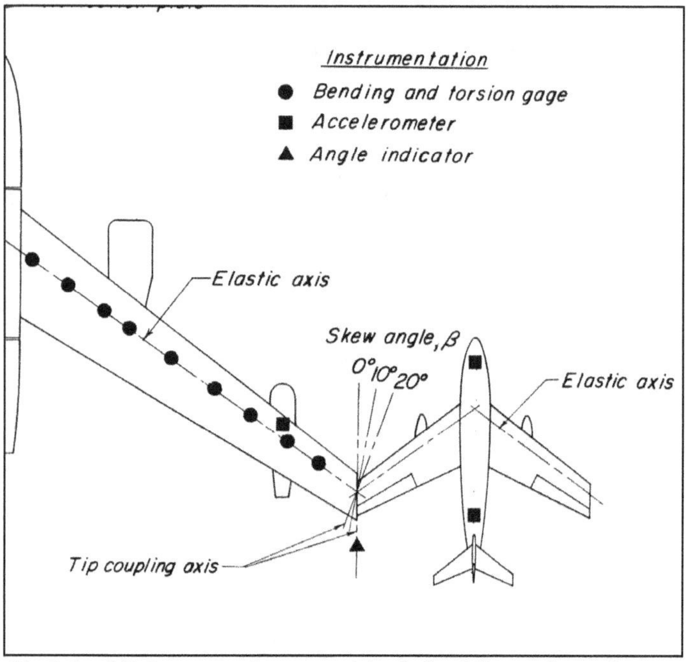


*Sketch of linkage system used to deflect the outboard aileron of fighter model in response to the bank angle of the fighter with respect to the bomber. (NACA RM152e08)*

The investigators found that moving the fighter models rearward with respect to the wing tip of the bomber improved the flying behavior of the composite model. In the rear position, the pitching motions of the fighters were reduced, which resulted in a corresponding reduction in the magnitude of their flapping motions.[v]

***June 1953, Floating Wing Tip Panel Flight Test Proposal***

The Wright Air Development Center proposed a demonstration of actual floating, wing tip panel flight hardware. A pair of OQ-3 target drones would be mounted on the wing tips of a Beech C-45G to simulate fuel-carrying, wing tip panels. The Frankfort Company of Frankfort, Indiana manufactured OQ-3 target drones in the closing years of World War II.

By the middle of 1953, the Wright Air Development Center was seeking a new prime contractor to replace Republic Aviation, because Republic had not performed assigned work in accordance with the contract.

***Simulated Flight Investigation of a Scaled-Speed Elastic Swept-Wing Bomber and Fighter Models Coupled Wing Tip to Wing Tip, February 1956***

Potential future application of wing tip coupling was expected to utilize a large jet propelled airplane such as the Boeing B-52 as the carrier. Large, swept wing jets are considerably more flexible than the Convair B-36. Beyond the basic problems of controlling the parasite while coupled to the wing tip, the flexibility of the carrier airplane would directly affect the flutter and stability characteristics of the composite aircraft.

Wind tunnel tests of wing tip coupling had used rigid models. There was no way to mathematically predict the behavior of a fighter connected by a hinge to the tip of a flexible wing. It was necessary to conduct wind tunnel tests of a wing tip coupled, composite aircraft that modeled the elastic and mass characteristics of the subject airplanes as well as their aerodynamic characteristics.

Investigators at the Langley Aeronautical Laboratory conducted tests of an elastically-scaled model of a swept wing bomber with a swept wing fighter coupled to its wing tip in the Langley 300-mile per hour, 7 by 10-foot wind tunnel in 1955. Robert F. Thompson was the principle investigator. A primary objective of the tests was the observation of the nature of the changes in the flying qualities of a flexible composite airplane at different air speeds. The team was expected to establish the highest feasible air speed at which a flexible composite aircraft could be flown.

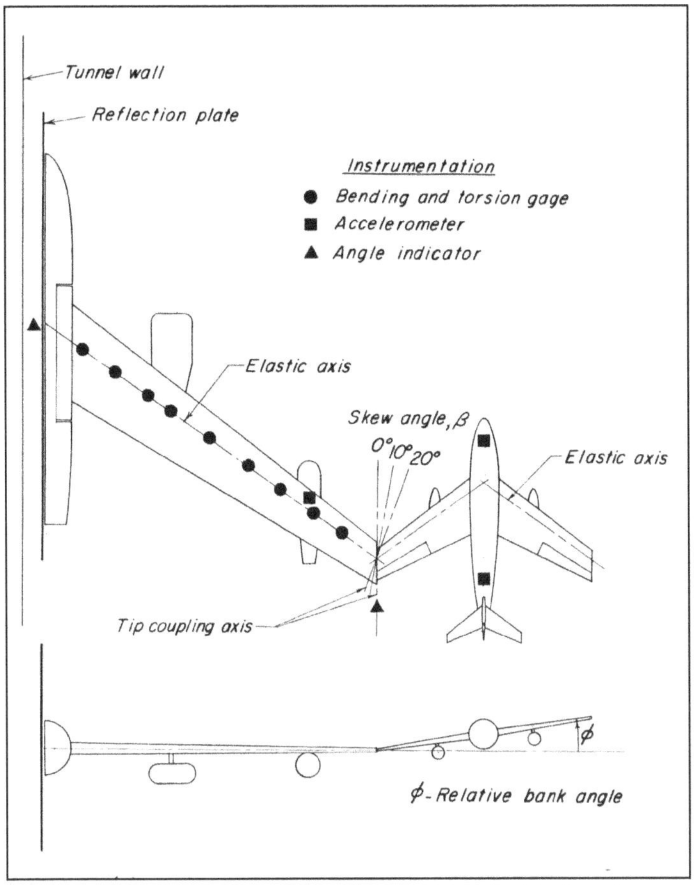


Sketch of test configurations. Fighter coupled to bomber, wing tip to wing tip. (NACA RM155j24)

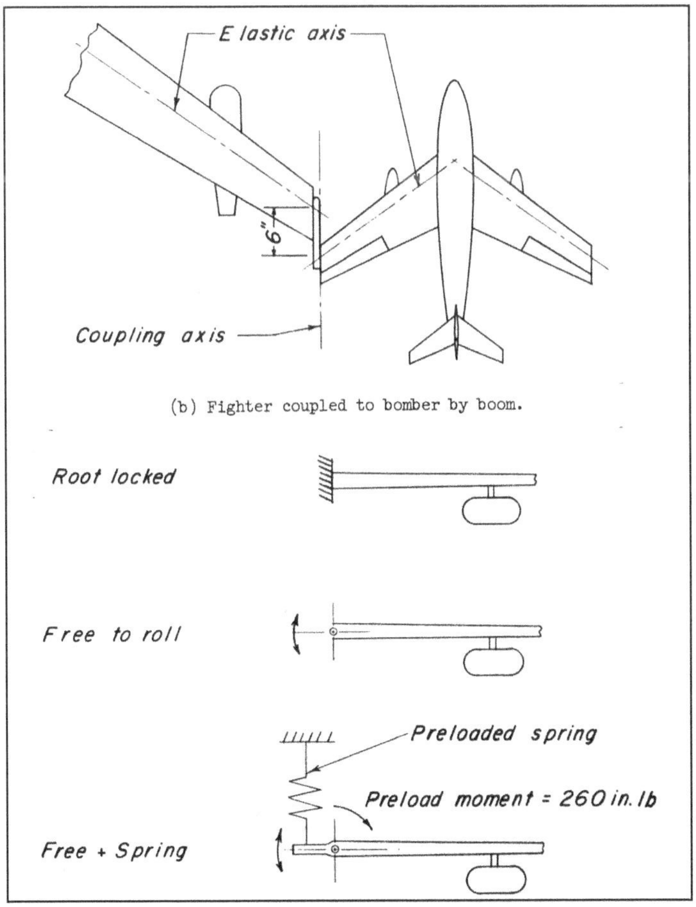


Sketch of test configurations. Bomber root conditions tested. (NACA RM155j24)

The bomber model was designed to be representative of a contemporary, operational, swept wing aircraft with very flexible wings. Although the model did not represent any particular actual airplane, it was considered to be a 1/14-scale model of a generic bomber. Aerodynamic scaling effects caused the model to perform like a full-scale airplane flying at 20,000 feet altitude.

Only half of the bomber and one of its wings were modeled for the tests. The bomber fuselage was cylindrical in cross section. The half span of the model was 49.4 inches. The wing was swept 34.5 degrees. It had two engine nacelles. The inboard engine nacelle represented a twin engine nacelle while the outboard nacelle represented a single engine nacelle.

The elastic properties of the wing and its mass distribution were accurately reproduced in scale. A duralumin spar was machined to have flexibility similar to a full scale wing spar. The cross-section of the spar was a cross shape. The spar design allowed the engineers to adjust the ratio of the stiffness of the spar in the up and down direction to its stiffness in the plane of the wing to accurately mimic a full-scale wing. The spar was about seven times as stiff in the plane of the wing as it was in the vertical plane.

Nineteen balsa wood airfoil segments were attached to the spar independently so that they would not affect the stiffness of the wing. Each segment had a duralumin rib in the center. Each rib had mounting points for small ballast weight in front of and behind the spar. Each ballast weight was selected to reproduce the distribution of mass in an actual wing. A narrow strip of sponge rubber was used to seal the gap between each pair of airfoil segments.

A full-span fighter model was hinged to the wing tip of the bomber model. It was 31.4 inches long and had a wingspan of 31.9 inches. The wing and tail were swept 35 degrees.

The main structural component of the fuselage was a steel spar that was essentially rigid. A balsa wood shell with hardwood mounting ribs formed the fuselage contour.

The construction of the wing of the fighter was similar to the bomber's wing in order to model the elastic characteristics of the full-scale airplane. Small pods were attached to the underside of the wing to simulate the mass of external fuel tanks. The pods were removed for one flight. The removal of the pods did not significantly affect the behavior of the models.

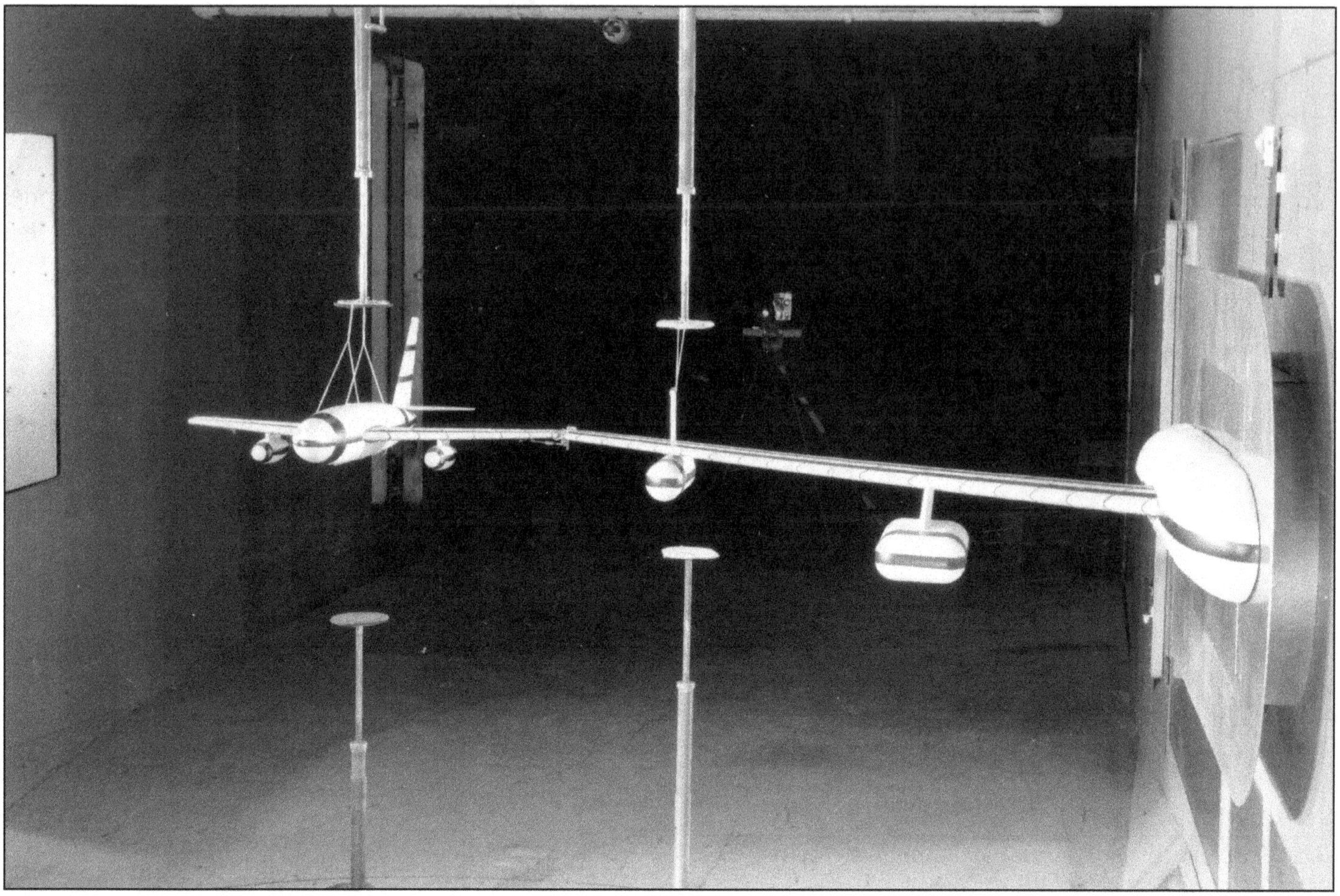

*Model as viewed from upstream in tunnel. (NASA Photo L-79952)*

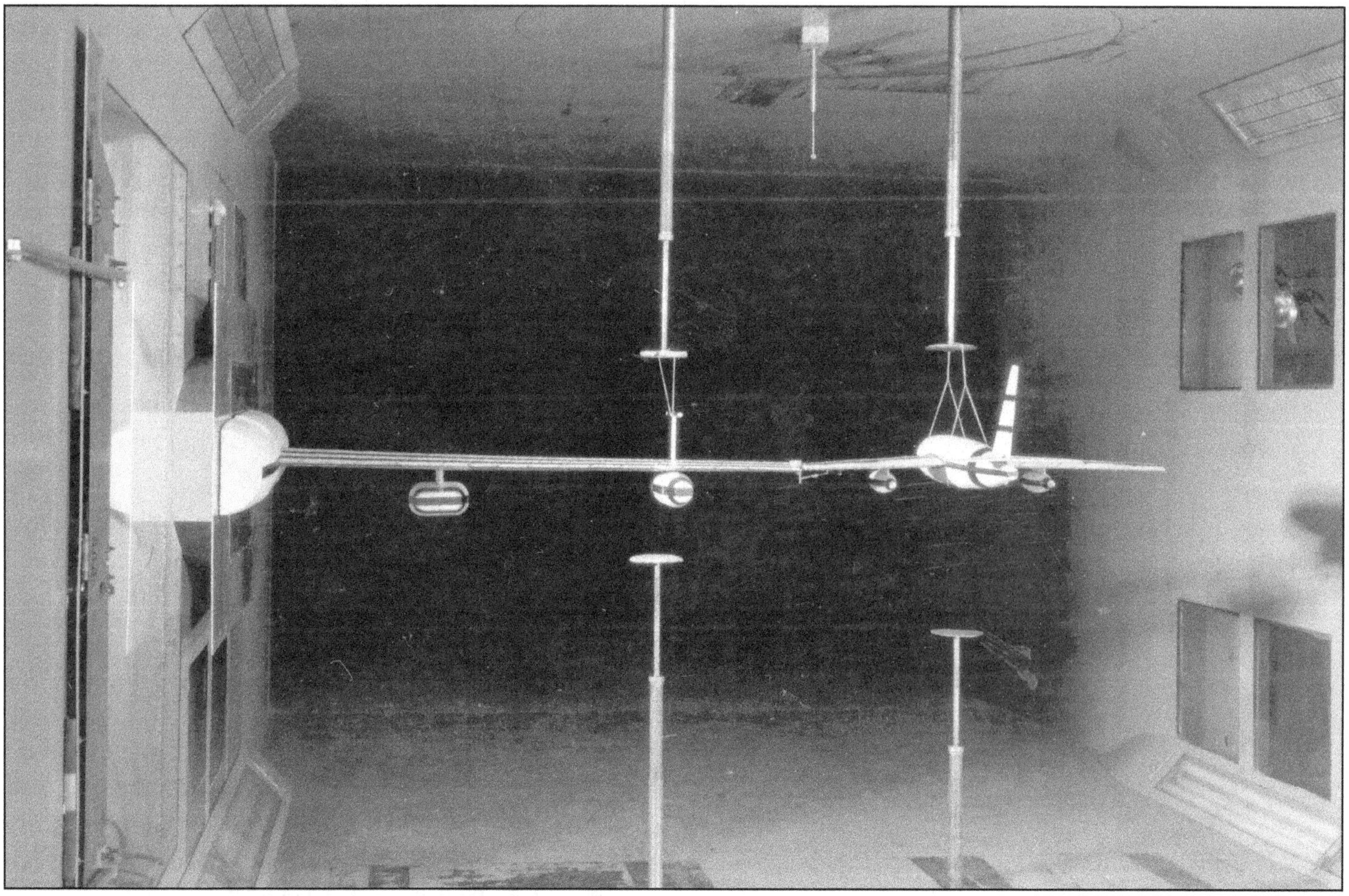

*Model as viewed from downstream in tunnel. (NASA Photo L-79951)*

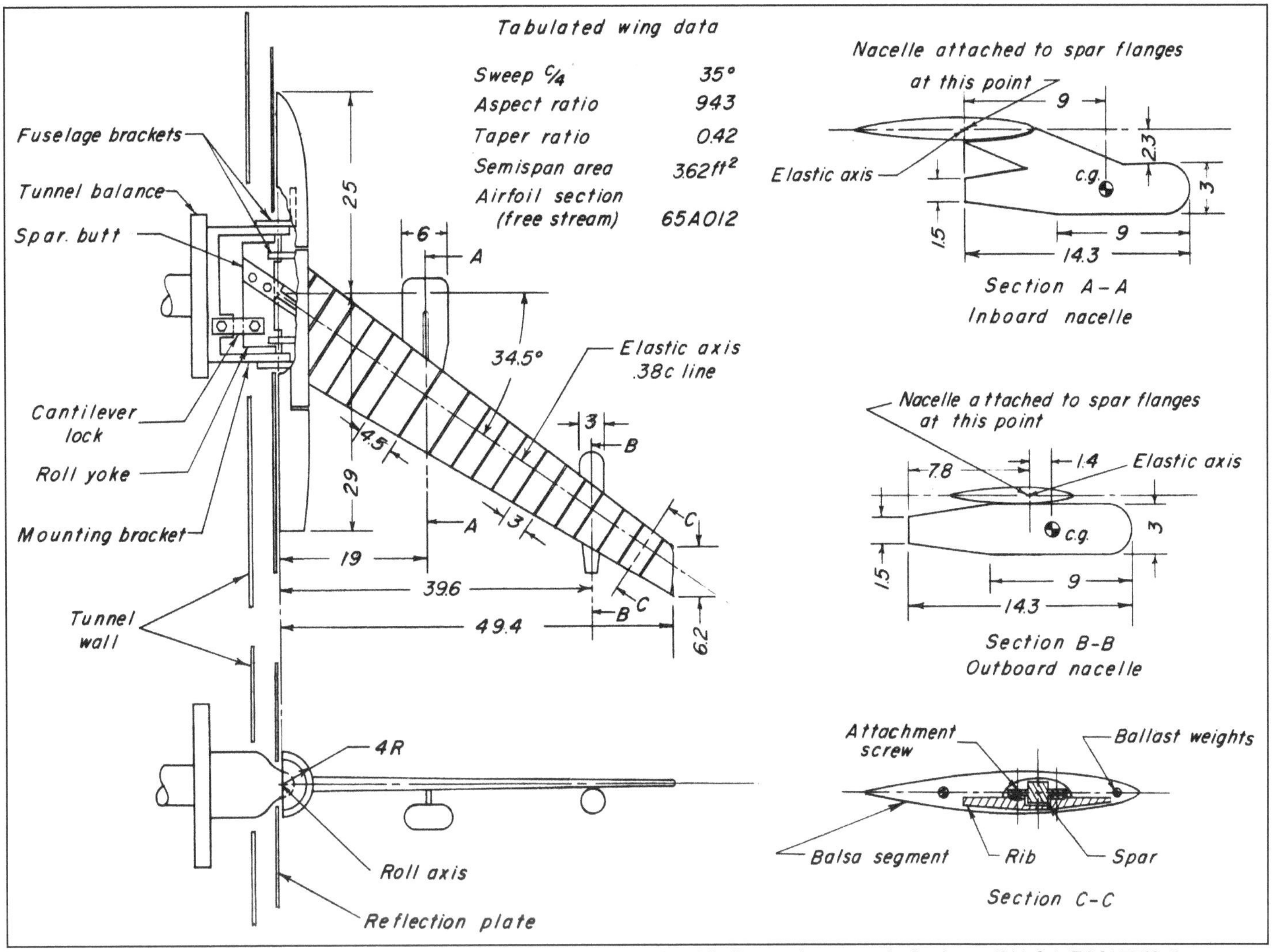

*Drawing of semispan bomber model and tunnel mounting details. All dimensions are in inches. (NACA RM155j24)*

The bomber model was tested separately in the wind tunnel. The investigators slowly increased the air speed in the wind tunnel to establish the speed at which the wing of the bomber would begin to flutter. The onset of flutter determined the maximum speed that the model could be flown. The onset of flutter occurred between 890 and 940 scale miles per hour depending on the test condition.

The fighter was fixed to the wing of the bomber in pitch and yaw. It was free to roll about the wing tip coupling.

As in the previous wing tip coupling wind tunnel tests, the fighter model was equipped with a linkage to adjust the position of the ailerons to compensate for the rolling movement of the fighter around the bomber's wing tip. The linkage of push-pull rods and bell cranks was contained entirely within the wing.

Unlike the previous wind tunnel tests, the ailerons were rigged conventionally. They moved in opposite directions. The inboard aileron operated to reduce the effectiveness of the outboard aileron at stabilizing the fighter, but the configuration was more representative of actual flight hardware.

The gearing ratio of the aileron linkage could be adjusted. The minimum gearing ratio moved the aileron through an angle 0.6 times the amount of the flapping angle. This was the same as the gearing ratio used in the full-scale B-29/F-84 wing tip coupling tests. The maximum gearing ratio moved the aileron through an angle 1.4 times the flapping angle.

The fighter model could be attached directly to the wing tip of the bomber or it could be attached to the bomber's wing tip by a boom that positioned it farther aft. This was done to determine whether the aft position would improve the flying qualities of the composite aircraft, although the rear placement of the fighter was expected to reduce the aerodynamic benefits of the wing tip coupling method.

The coupling mechanism on the bomber wing tip could be adjusted to change the relative angle of attack of the fighter and bomber. By setting the relative angle of attack the investigators controlled the amount of lift

*Model rigged for still-air vibration survey. (a) View from upstream in tunnel. (NASA Photo L-79663)*

generated by the wing of the fighter and its flapping angle.

The coupling mechanism could also be adjusted to skew its forward end outboard relative to the rear end. This provided a restoring force when the fighter rolled away from the wings-level position.

The bomber model was fixed to the wall of the wind tunnel in pitch and yaw. For some tests it was also fixed in roll. In others, it allowed to roll around its longitudinal axis. The bomber was fixed in position to represent conditions of symmetric load. If both wings of the bomber were supporting an equal load, the mid-plane of the bomber would remain essentially fixed in position.

The longitudinal axis of the bomber was unlocked to allow the model to roll for investigations of asymmetric load conditions. The wing spar of the bomber extended past the middle of the fuselage of the model. A spring was attached to the top of the spar to simulate the lift generated by the opposite wing.

The tests showed that it made little difference to the flying qualities of the model whether the fuselage of the bomber was allowed freedom of movement about the roll axis or was fixed in position.

The center of lift of the fighter was well forward of the wing tip coupling. This increased the amount of twisting force that the fighter applied to the bomber's wing. Any change in the pitch angle of the fighter model twisted the bomber model's wing. The twisting of the bomber's wing increased the pitch angle of the fighter. The increase in the pitch angle of the fighter applied more force to the bomber's wing, causing it to twist further. Above a critical air speed, the value of the twisting force would rapidly increase. In engineering terms, the wing would diverge under torsion. The twisting force imposed by the fighter would quickly exceed the strength of the bomber's wing. It was the same sort of divergence under torsion that made the manufacture of a high-speed, forward swept wing airplane unfeasible until composite materials of sufficient strength were developed.

The tests demonstrated that the flexible composite aircraft had satisfactory flying qualities up to a scale air speed of 400 miles per hour. At higher air speeds, the composite aircraft began to display divergent behavior. The critical divergence speed of the composite aircraft was well below the limiting flutter speed of the bomber model flying independently.

The onset of divergence could be extremely abrupt. There was a significant danger of tearing the fighter model from the wing of the bomber as the air speed in the wind tunnel was increased. A large flat plate was mounted on hinges downwind of the models. If the model started to diverge, the plate would be released to obstruct the airflow through the tunnel and rapidly reduce the air speed through the test section.

Adjusting the gearing ratio of the aileron linkage had no measurable effect on the critical speed for

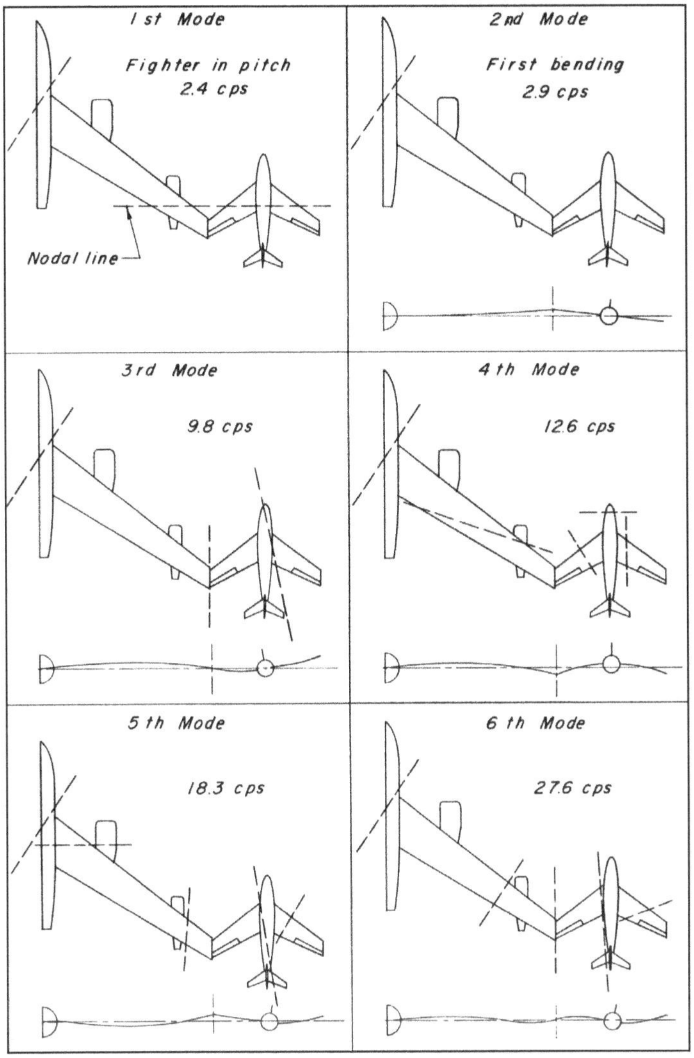


*Still-air natural vibration modes of test model. Models coupled, wing tip to wing tip. Bomber root locked. (NACA RM155j24)*

divergence. The fighter model stabilized more rapidly after disturbances at the higher gearing ratios.

The models were designed to investigate the effect of a skewed wing tip coupling as a purely aerodynamic means of stabilizing the fighter on the bomber's wing tip. Skewing the wing tip coupling 10 degrees lightly damped the flapping of the fighter model.

Increasing the skew angle to 20 degrees produced an extremely undesirable flapping of the fighter at air speeds well below the critical divergence speed. When the bomber model was allowed to rotate about its roll axis, the fighter would roll up and down approximately six degrees. The initial frequency of the oscillation was about 1.5 cycles per second. As the air speed increased, the magnitude of the oscillations remained about the same, but the frequency of the oscillations increased to about 1.8 cycles per second. When the fuselage of the bomber was fixed in all three axes, the magnitude of the flapping of the fighter would vary between four degrees and fifteen degrees as its oscillations moved in and out

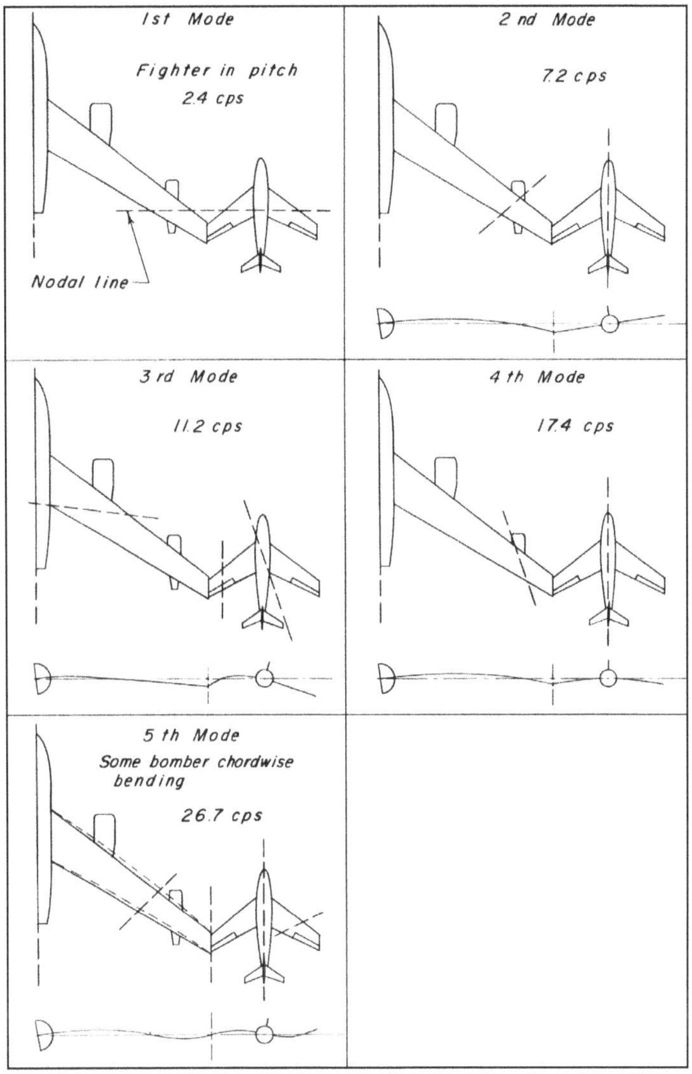


*Models coupled, wing tip to wing tip. Bomber free to roll and root free to spring. (NACA RM155j24)*

of phase with the flexing of the wing of the bomber. The magnitude of the flapping of the fighter at the 20-degree skew angle was reduced somewhat, but not eliminated, by increasing the gearing ratio of the ailerons.

Mounting the fighter model on a boom to place it farther aft relative to the wing tip of the bomber would reduce the twisting force that the fighter imposed on the wing of the bomber. It was hoped that this would increase the speed at which divergence occurred. The rear position had no appreciable effect on the speed of the onset of divergence.[vi]

#### *Flutter Tests of a 1/25-Scale Model of the B-36J/RF-84F Tip-Coupled Airplane Configuration in the Langley 19-Foot Pressure Tunnel, March 1956*

The hinge axis of the wing tip-coupling mechanism of the GRB-36F was skewed with its forward end farther outboard than the rear end. This provided a restoring force when the wing tip-coupled RF-84F rolled away from the wings-level position. When the Thunderflash rolled upward, its nose would pitch downward. The increased pressure on the top of the wing would push it back down. When the Thunderflash rolled downward, its nose would pitch up, providing an upward restoring force. As the airspeed of the wing tip-coupled aircraft is increased, the magnitude of the restoring-force increases.

Above a critical airspeed, a condition called flutter sets in. The wing tip-coupled fighter will begin flapping up and down. The direction of rotation of the vortex streaming from the fighter's free wing tip will reverse each time the wing passes through the position where its angle of attack is zero. The strength of the vortex will increase directly with the angle of deflection of the fighter. This pumps energy into the flapping of the RF-84F, so the angle of deflection gets greater with each oscillation. Unless the airplanes immediately slow down or the fighter is jettisoned, the divergent flapping will eventually cause a failure of the wing tip-coupling mechanism or the wing of one or both airplanes.

Of course, the flutter characteristics of the wing tip-coupled B-36 and RF-84F were of major concern to all aboard the aircraft. Tests of a 1/25th-scale model of the wing tip-coupled B-36 and RF-84Fs were performed in the 19-foot pressure tunnel at the National Advisory Committee for Aeronautics Langley Aeronautical Laboratory.

A primary objective of the tests was to determine the value of the flutter speed for different skew angles of the wing tip-coupling mechanism. Among other things, the flutter speed depends on the skew angle of the hinge. A larger skew angle will result in a lower flutter speed.

The tests also examined the behavior of the coupled airplanes below flutter speed, particularly with regard to the movement of the fighter relative to the bomber wing.

The Fort Worth Division of Convair provided the model of the B-36J. The wingspan of the B-36 model was nearly ten feet. It was not merely an accurate model of the shape of the B-36. Its mass and inertia were properly scaled as well. The modelers did not just produce a model that weighed the proper scale amount. They reproduced the distribution of the mass within the airplane. Each wing incorporated three weights to represent the concentrations of mass of the Wright 4360 radial engines.

The model was also designed to properly represent the elastic characteristics of a B-36. Metal spar structures modeled the elastic characteristics of the wing and fuselage. Weights attached to the spar structure represented the concentrations of mass in the actual airplane. The weights and structure were housed in balsa wood box structures that defined the aerodynamic form of the airplane. The gaps between the balsa sections were spanned with thin rubber sheets.

Between wind tunnel runs, the ailerons and rudder of the B-36J model could be manually adjusted to trim the model in roll and yaw. The horizontal stabilizer was remotely controlled to trim the model in pitch while the wind tunnel was operating. Part way through the test series, a remotely controlled flap was added to the right wing so the model could also be trimmed in roll while the tunnel was operating.

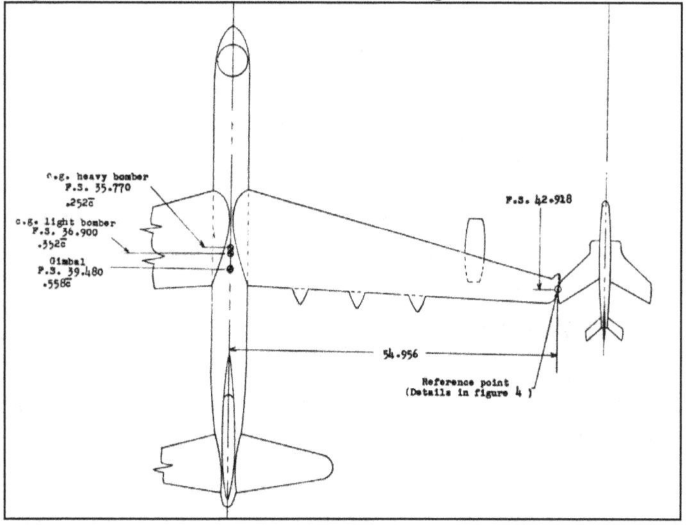


*General arrangement of model. (NACA RM SL56A25b)*

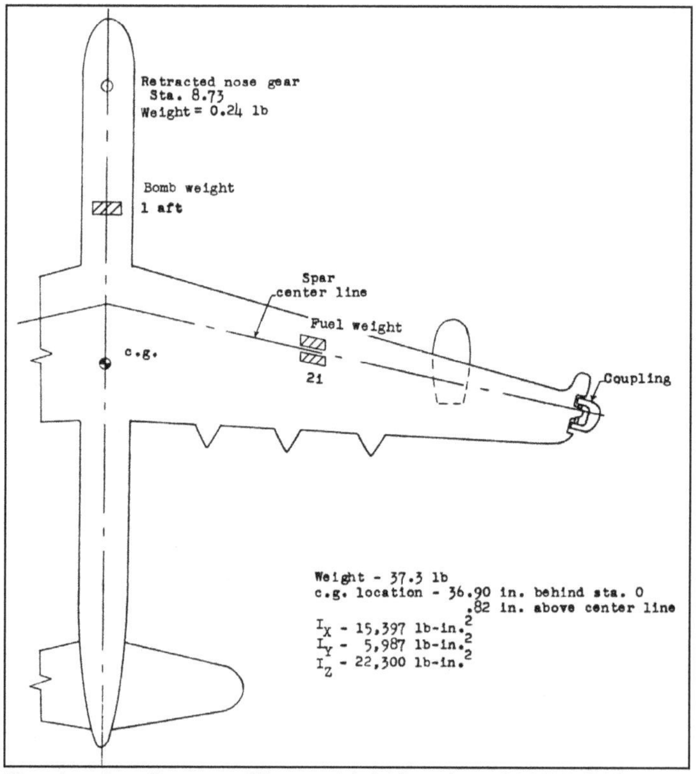


*Bomber loading conditions - Light bomber (NACA RM SL56A25b)*

*Fuselage of the 1/25th-scale B-36 model. (NASA Photo L-92441)*

The Thieblot Aircraft Company of Hagerstown, Maryland manufactured two 1/25th-scale RF-84F models. Like the B-36J model, they were scaled for mass and inertia. The elasticity of the fighters was negligible for the purposes of the tests, so the models were rigidly constructed.

The RF-84F models were attached to the wing tips of the B-36J model with hinges that were skewed outward at the front end. The skew angle could be set to 8 degrees or 15 degrees. Each hinge was pivoted at the front end so that the pitch angle of the hinge could be adjusted by remote control. Direct control of the pitch of the model gave a degree of control over its angle of bank.

The bomber model was supported in the wind tunnel by a vertical rod. During the first tests, a gimbal mechanism inside the fuselage of the model allowed it to rotate freely around all three axes. It was also free to translate up and down the support rod. Cables attached to the top and bottom of the fuselage controlled its position on the support rod. Cables were also attached to the bottom of each wing of the model to restrict its range of motion in roll.

The gimbal mechanism of the large and flexible model was mounted slightly behind its center of gravity. As a result, the bomber mount was unstable in pitch. The model had a strong tendency to move to the upper or lower stop on the mounting rod. To prevent damage to the model, during the latter stages of the tests it was restrained from rotation in pitch and yaw and fixed in place on the mounting rod. In this "fixed mode", it was only free to move around the roll axis.

The operating pressure of the 19-foot wind tunnel could be adjusted to simulate flight at different altitudes. For the flutter evaluation, the tunnel was pressurized to simulate flight at 25,000 feet altitude.

The actual airspeed through the wind tunnel was varied from 65 miles per hour to above 110 miles per hour.

Once the airspeed in the wind tunnel stabilized, the elevator of the bomber model was adjusted to make it fly level. The pitch adjustment of the wing tip couplings was adjusted to make the fighter models fly level with the wing of the bomber model. With the model trimmed out, the fuselage cables were jerked to start the model oscillating up and down. The test observers would note the period of the oscillation and time how long it took to stop. It was noted that random disturbances in the wind tunnel kept the fighter models from settling down completely.

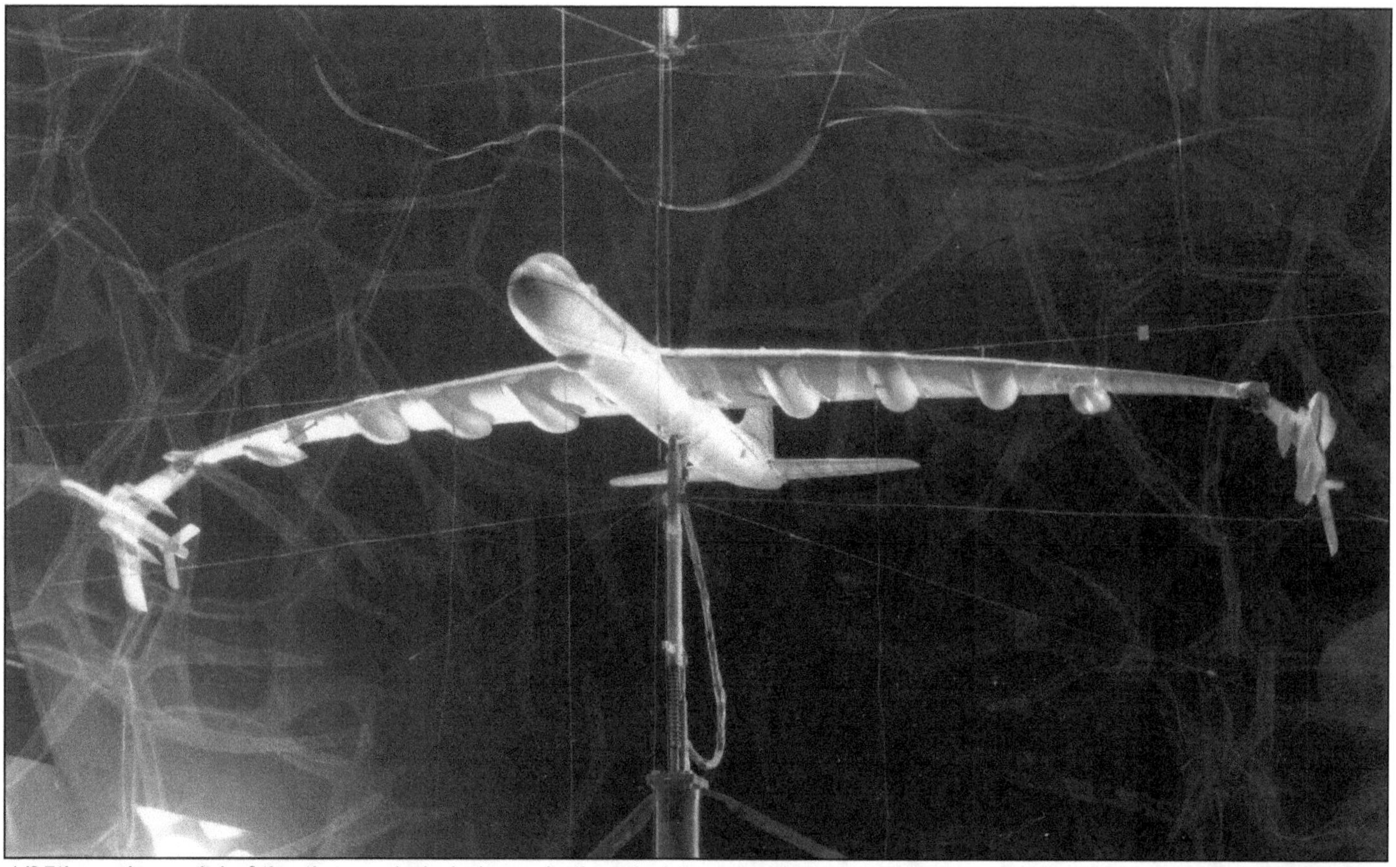

1/25th-scale model of the tip coupled airplanes in the wind tunnel at Langley Research Center. (NASA Photo L-88494)

1/25th-scale model of the tip coupled airplanes in the wind tunnel at Langley Research Center. (NASA Photo L-88498)

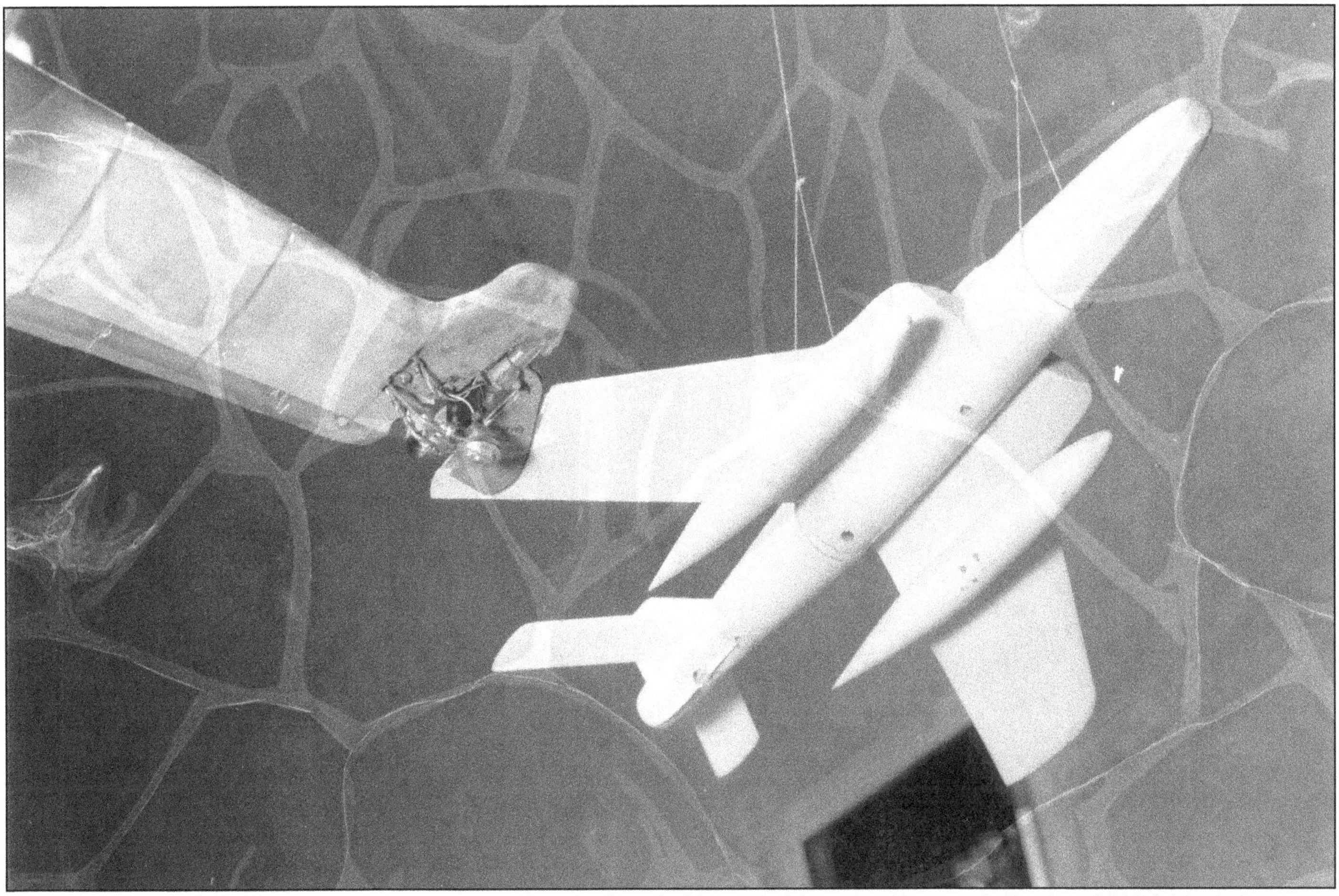

*RF-84F model and adjustable coupling mechanism in the wind tunnel at Langley Research Center. (NASA Photo L-88498)*

When the models had stabilized, one of the wing cables was jerked to roll the model momentarily so that its response could be observed. The airspeed was then increased to the next test value. The procedure was repeated until the airspeed exceeded the flutter speed for that configuration. When the fighter models began to flutter, the tunnel-fan drive motor was shut off and the test of that configuration was concluded.

Tests were run with the fighters attached in two different positions, at two skew angles, and at two different weights. The bomber was tested at two weights. Flutter evaluations for sixteen configurations were performed.

It was found that when the flutter speed was exceeded, the fighter models fluttered symmetrically. The simulated lightweight fighters would start rolling up and down as the speed approached the flutter speed. The amplitude of the flapping would slowly increase. The wing of the bomber model twisted back and forth with the rolling of the fighter.

The onset of flutter of the simulated heavyweight fighters was much more abrupt and violent. They immediately began flapping wildly, bending the wing tip of the B-36J model up and down.

Changing the skew angle of the coupling mechanism had a pronounced effect on the flutter speed. The flutter speed with the skew angle at 8 degrees was 30 percent greater than the flutter speed at a skew angle of 15 degrees.

It was found that moving the fighter models farther back on the coupling also increased the flutter speed.

A set of tests was run with just one fighter model attached to the bomber. The flutter speed was significantly higher than with both fighters attached.

When the bomber was yawed, the effective skew angle of each fighter's coupling mechanism was altered, one increasing, the other decreasing. The fighters would oscillate on the wing tips with different periods. The effective skew angle of the fighter on the trailing wing would increase and its flutter speed would decrease.

A small wing section was installed in the wind tunnel about 4 inches below and 1/2 inch in front of the model fighter on the left wing. It was used to simulate wind gusts by deflecting it momentarily. The resulting oscillation of the fighter diminished rapidly. The simulated lightweight fighter was deflected more than

*An airfoil mounted upstream in the wind tunnel was used to simulate wind gust loads. (NASA Photo L-89576)*

*An airfoil mounted upstream in the wind tunnel was used to simulate wind gust loads. (NASA Photo L-89576)*

the heavyweight fighter was. The fighter was deflected more at the 8-degree skew angle than at the 15-degree skew angle.

With the bomber fixed except for the roll axis, the fighters were trimmed at various bank angles from 24 degrees down to 30 degrees up. When the fighters were banked up, the combination would rock its wings. The amount of rocking would slowly increase over time. The combination flew steady in a test with the bomber free to move and the fighters trimmed so that they were banked 24 degrees down.

The Langley Aeronautical Laboratory report was published on January 11, 1956, shortly before the Tom-tom flight-test project was begun. [vii]

### *Project Long-Tom*

Project "Long-Tom" was an investigation of the behavior of fuel-carrying, free-floating panels using actual flight hardware for comparison with the results of the wind tunnel tests. The chief test pilot for the project was S. C. Tuttle. T. Gillespie assisted Tuttle on the test flights. The tests were conducted at Beechcraft's pant in Wichita, Kansas.

Under Air Force Contract AF33(600)-27462, Beechcraft modified a standard L-23, 55-3465 with a pair of hinged wing panels. The L-23 was the military equivalent of the civil Beechcraft Model 400 Twin Bonanza. The modified airplane was designated XL-23C. The aspect ratio of the basic Twin Bonanza was 7.5. The addition of the wing tip panels gave the XL-23C an aspect ratio of 13.4.

The panels were held against the wing tips of the XL-23C by a pair of hooks from F-84 Thunderjet drop tanks. Each panel pivoted against a pair of ball and socket joints mounted on either side of the drop-tank hook. This arrangement allowed the panel to rotate around the end of the wing, but fixed the angle of attack of the panel relative to the wing of the Twin Bonanza. Beechcraft engineers used the term floating angle to denote the angle of deflection of the wing tip panel relative to the wing.

Each panel had a trailing edge flap. The flaps were geared to move in response to changes in the floating angle of the panels. The pilot of the XL-23C could vary the gear ratio of the flap movement in flight. The gear ratio could be adjusted from 1½ to 4 times the floating angle.

Each panel carried a torpedo-shaped structure that represented a fuel tank. Each torpedo carried ballast in its nose to prevent flutter. Each torpedo had a fixed vertical stabilizer and a movable horizontal stabilizer. The movement of the horizontal stabilizer was connected to the deflection of the trailing edge flap on the panel. By operating together, the flap and stabilizer adjusted the lift of the panel without changing its pitch angle. This eliminated the twisting force that EF-84D Thunderjets had applied to the wing of the EB-29B Superfortress.

Each panel was equipped with a single-wheel

*Beechcraft XL-23C 55-3465 with a pair of hinged wing panels. (Bud Anderson Collection)*

*Beechcraft XL-23C 55-3465 flying with a single hinged wing panel. (Bud Anderson Collection)*

landing gear. The landing gear strut retracted into the torpedo. They did not swivel, but that did not affect the turning radius of the XL-23C. In very sharp turns the inboard wheel actually moved backwards.

A micro switch was set to detect when the panel exceeded a floating angle of plus or minus 30 degrees. If the floating angle became excessive, the micro switch triggered the release the wing tip panel. The pilot of the XL-23C was provided with an electrically operated jettison button in the cockpit. There was also a manual release provided for use in the event of a failure of the electrical release.

A spring-loaded spoiler was mounted on the outboard span of the panel. As the panel was jettisoned, a cable would release the spoiler to insure that the panel would roll away from the Twin Bonanza.

The wing tip panel assemblies were given high visibility markings to assist in their recovery following jettison. Wide black stripes were painted around the torpedoes and a checkerboard pattern covered their lifting surfaces.

The wing tip panels angled downward from the wing tip of the Twin Bonanza by several degrees when their landing gear wheels rested on the ground.

The behavior of the modified Twin Bonanza during take-off remained essentially unchanged. The pilots could effectively ignore the wing tip panels. The downward angle of the wing tip panels caused their flaps and horizontal stabilizers to deflect downward during the take-off run, increasing the lift of the panels. The panels raised off of their landing gear at a speed well below the take-off speed of the Twin Bonanza. The gear ratio of the panel trailing-edge flap displacement would be adjusted before take-off to compensate for the quantity of fuel in the tanks.

After take-off the pilot of the XL-23C adjusted the trim of the panels so that they lined up with the wings at cruising speed.

As the angle of attack of the Twin Bonanza changed, the floating angle of the panels also varied. As the angle of attack decreased at higher speeds, the panels floated downward slightly. As the angle of

*Beechcraft XL-23C 55-3465 with a pair of hinged wing panels landing at Wichita, Kansas. (Bud Anderson Collection)*

attack increased at lower speeds, the panels floated up slightly. The panels floated upward in turns for the same reason. The floating angle increased with the amount of g-force. The floating angles experienced were small in comparison with the angle that would trigger the automatic jettison mechanism.

The Twin Bonanza pitched forward slightly as the pilot lowered its landing flaps, reducing the angle of attack of the airplane. The sudden reduction in angle of attack caused the wing tip panels to drop sharply relative to the wing. If the pilot reduced the angle of attack still further by applying power without adjusting the trim of the panels, it was possible for the wings to float downward far enough to trigger the automatic jettison mechanism.

The XL-23C did not suffer any significant longitudinal stability or control problems. Pitch control was responsive.

The directional stability and control of the XL-23C were less satisfactory. The rate of roll was reduced slightly, but was still satisfactory. The addition of the wing tip panels actually improved the efficiency of the ailerons slightly by eliminating the wing tip vortices that normally existed in their vicinity.

As the airplane was rolled, the panel in the wing that was moving up would float slightly downward while the panel on the wing moving down would float slightly upward.

As an airplane rolls, the wing going up produces more drag as well as more lift. This causes the airplane to yaw in the opposite direction of the intended turn. This effect is called adverse yaw. The adverse yaw of the XL-23C was higher than usual. A large rudder pedal deflection was required to fly a properly coordinated turn.

The damping of panel oscillations varied with the trailing edge flap gear ratio. At low air speed and a low gear ratio the panels stabilized quickly. While cruising at a speed of 120 knots with a high flap gear ratio it was possible to start an oscillation of the panels that did not die out. The pilot could stop the oscillation by reducing the gear ratio or by increasing the angle of attack by slowing down or pulling up.

The stall behavior of the XL-23C was good. The wing of the Twin Bonanza incorporated 4 degrees of washout. That means that the end of the wing has an angle of attack 4 degrees less than the wing root. This insures that the wing root stalls before the wing tips, reducing the tendency for the airplane to drop one wing during a stall. The washout also insured that the wing tip panels did not stall before the wing itself.

The wing loading of the XL-23C was not increased, so its landing and take-off speeds were the same as those for the basic Twin Bonanza. Its take-off roll was somewhat longer because the extra weight reduced its acceleration. The maximum speed of the XL-23C was reduced slightly by the drag of the panels. Its rate of climb was reduced by the extra weight. The improved aerodynamic efficiency that resulted from the increase in aspect ratio provided by the wing tip panels actually increased the service ceiling.

The fuel that could be carried in the panels tripled the range of the airplane with only minor reductions in its basic performance.

Investigators suggested that a more fully automated control system might be developed. It was felt that the stabilizing effect of the trailing edge flaps and horizontal stabilizers could be improved by installing a variable ratio deflection system with a low gear ratio at small floating angles and increasing gear ratios at higher floating angles.

Another potential improvement would be the installation of a system that adjusted the trim of the wing tip panels as the air speed of the XL-23C varied. [viii]

## MX-1018 Tom-Tom, 1953 - October 1956

### Project Initiation

Plans for the continued development of the wing tip coupling concept were already in place when the MX-1016 tests were brought to an untimely conclusion by the crash of the EB-29B and EF-84D in March 1953.

Further development of the wing tip coupling method was conducted under Air Force Project MX-1018. The Convair GRB-36F 49-2707 that had been Project FICON prototype carrier was selected to be the mothership for the program. As in the FICON project, the Republic RF-84F was selected for the parasite airplane. In the early 1950s the mission of the parasites had shifted from providing fighter escort for bombers on penetration missions to conducting independent reconnaissance missions into areas that the reconnaissance-bomber could not safely penetrate.

Project Tom-Tom would also evaluate a proposed range extension for the B-36 by the addition of "floating" fuel-bearing wing tip panels. A B-36 equipped with such panels would have a wingspan exceeding 300 feet.

Major General Thomas Gerrity was the USAF project officer for MX-1018. Thomas Sullivan was the Convair, Fort Worth Division Contracts Supervision administrator for the project. They coined the security code Tom-Tom for the project.

Project Tom-Tom was conducted at the Convair plant across the runway from Carswell Air Force Base. FICON and Tom-Tom were handled through a small group in the B-36 Project Office. Beryl Erickson was the Manager of Flight Operations and the Chief Test Pilot. Major Henry Duchendorff was the Commander of Flight Operations for the Air Force at Air Force plant No. 4.

Two German expatriates who were assigned to the Pentagon participated closely in engineering decisions about the project. The two individuals rode along on every flight. Beryl Erickson described the German engineers as "prone to quietness".

There were sixteen pilots assigned to the B-36 Project Office. Erickson assigned four pilots to the FICON and Tom-Tom projects: Raymond Fitzgerald, A.S. "Doc" Witchell, George Davis, and himself. Erickson and Fitzgerald agreed to alternate piloting the GRB-36F and the RF-84F.[ix]

### Advantages

There were advantages to be gained by attaching a Republic RF-84F to each wing tip of a B-36 rather than suspend one from a trapeze in the bomb bay. The improved aerodynamic efficiency gained from the increased aspect ratio of the wings offset much of the drag of the fighters Without a parasite occupying its bomb bay, the RB-36 carrier could be used to conduct a separate bombing mission.

Having two parasites aboard the carrier added another dimension to mission planning. The second parasite could act as a fighter escort for the primary reconnaissance parasite. In the event that the primary airplane failed to reach its target, the second parasite could take over the reconnaissance mission. On a nuclear strike mission, the second parasite could act as a backup strike aircraft in the event that the primary strike aircraft failed to deliver its weapon to the target. If the primary strike was successful, the backup Thunderflash could strike a target of opportunity on the way to the retrieval point.

A significant disadvantage of the Tom-Tom mission concept was that it required each parasite pilot to remain in the cockpit of his Thunderflash for the duration of the RB-36 carrier airplane's mission. He would not be able to stand up or even stretch his legs out straight for 30 to 40 hours. He would have to eat, sleep, and relieve himself sitting in the confines of his cockpit, yet he would have to be physically and mentally capable of conducting his own penetration mission twenty hours or more after take-off.

#### *Wing tip coupled B-47*

Consideration was even given to the concept of coupling a pair of Boeing B-47 Stratojets onto the wing tips of a B-36. Evidence for this is found in a NACA Research Memorandum written at the Langley Aeronautical Laboratory entitled, *Calculated Lift Distributions of a Consolidated Vultee B-36 and two Boeing B-47 Airplanes Coupled at the Wing Tips.*[x]

If the feat had ever been attempted, the wingspan of that monster would have exceeded 460 feet. It would have had sixteen jet engines hanging from the wings in addition to the six reciprocating engines in the wings of the B-36. There would have been six bomb bays for nuclear weapons.

#### *Initial Tests*

For the initial evaluation of the Tom-Tom concept, an unmodified RF-84F Thunderflash was flown in tight formation with the wing tip of a standard B-36 in 1952. The pilot evaluated the strength of the vortex streaming from the wing of the B-36 and its effect on the behavior of the RF-84F.

Erickson conducted an informal evaluation of the view angles of the RF-84F pilot. The Thunderflash pilot had to look back over his shoulder to see the wing tip of his airplane while attempting to make contact with the wing tip of the JRB-36F. It was a very uncomfortable position. With the throttle in the pilot's left hand and the control stick in his right hand, his ability to look over his right shoulder was restricted to the point that it was far more difficult to engage the coupling mechanism on the left wing tip of the JRB-36F than the right wing tip.

#### *August 27, 1953 Initial Proposal*

The Air Force presented its proposal for the B-36 wing tip coupling experiments to Convair on August 27, 1953.

#### *June 1954, GRB-36F Transferred to Tom-Tom*

GRB-36F 49-2707 concluded its participation in Project FICON in June 1954. It was no longer suitable for operational missions, and it was not economically practical to return the GRB-36F to stock configuration, so it was transferred to project Tom-Tom. Convair was awarded Air Force Contract AF 33 (600) 23415 to modify the GRB-36F for the wing tip-coupling experiments.

Convair technical design personnel were already heavily involved in the development of the F-106A Delta Dart and B-59A Hustler, so they subcontracted the engineering study and design of the wing tip couplings to the Thieblot Aircraft Company. Convair fabricated and installed the wing tip coupling mechanisms into the test airplanes.

#### *August 26-29, 1954 Wing Tip Proximity Flights*

Convair pilots evaluated the behavior of a Republic RF-84F flying near the wing tip of the GRB-36F at Fort Worth on August 26, 27, and 29, 1954. The preliminary evaluation suggested that it was feasible to hook a Thunderflash to the wing tip of the GRB-36F.

#### *November 12, 1954, Development Engineering Inspection of Project Tom-Tom*

The Thieblot Aircraft Company hosted a Development Engineering Inspection of the Project Tom-Tom design on November 16 and 17, 1954. On the first day of the conference, the attendees were treated to a series of movies depicting wind tunnel tests, flights of an RF-84F in the proximity of a B-36 wing tip, and wing tip coupling tests conducted with the EB-29B and EF-84Ds. Company engineers delivered short talks covering the planned development phases of the Tom-Tom Project. Attendees were shown a full-scale

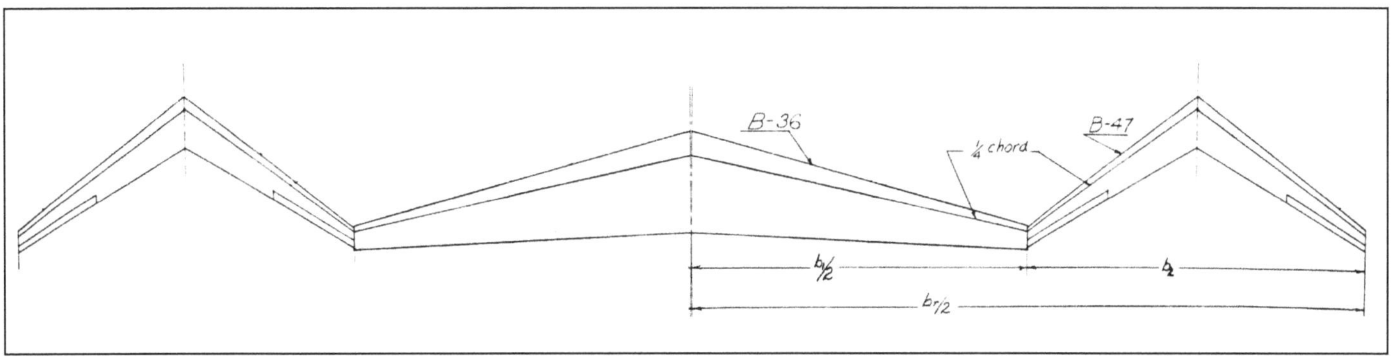


*Planform of the wing of a B-36 with a pair of Boeing B-47 Strataojets coupled to its wing tips. (NACA RM 150i26)*

The lightweight wing tip coupling mechanism installed on the left wing tip of the GRB-36F for the initial phase of Project Tom-Tom. (Lockheed-Martin Tactical Aircraft Systems Photo 26-1807)

The lightweight wing tip coupling mechanism installed on the right wing tip of RF-84F 51-1848 for the initial phase of Project Tom-Tom. (LMTAS Photo 26-1804)

mockup of the wing tip coupling mechanism that would be installed on the GRB-36F. On the second day of the conference, the engineers involved in the project met for group discussions of the various phases of project development.[xi]

### Modifications to GRB-36F

The Tom-Tom flight test program was conducted in two phases. During the first phase in 1955, a wing tip-coupling mechanism was installed on only the left wing of the GRB-36F, 49-2707. The mechanism was intended to test just the approach and initial contact between the RF-84F and the GRB-36F. It did not incorporate a locking mechanism to secure the parasite fighter to the bomber's wing tip. There was no plan to retract the capture mechanism while the RF-84F was attached during the first phase of the tests. For the second phase of the tests, it was planned that fully functional wing tip coupling mechanisms would be installed on both wing tips of the GRB-36F.

The wing tip coupling receiver consisted of a lightweight scissors structure bolted through the wing tip of the GRB-36F. The planform of the wing tip was not modified. The scissors unit was composed of a pair of arms that were connected to the wing tip of the GRB-36F at the rear end. The forward end of the unit pivoted outward from the wing tip. Wire control cables extended from the operator's station to the wing tip to operate the mechanism.

A receiver plate was mounted horizontally at the forward end of the scissors unit. The receiver plate was pivoted to the two arms so that its orientation did not change as the scissors unit moved in or out. A sheet metal aerodynamic fairing enclosed the scissors unit and receiver plate in the retracted position. The scissors mechanism was painted red-orange.

Despite its retirement from the FICON project, the bomb bay doors were not re-installed in 49-2707.

### July - August 1954, Delivery of Thunderflashes

Two Republic RF-84F Thunderflashes were assigned to the Project Tom-Tom. RF-84F 51-1848 was delivered to the Air Force on July 28, 1954 and RF-84F 51-1849 was delivered on September 27, 1954. They followed the prototype FICON RF-84F 51-1847 off the Republic production line. The Air Force handed them over to Convair in Fort Worth to be modified for the Tom-Tom project.

A wing tip coupling mechanism was installed on the right wing of RF-84F 51-1848. The initial test installation on 51-1848 was a jaw-shaped capture mechanism. It did not incorporate a rear pitch lock. The jaw capture mechanism fit around the leading edge of the wing of the RF-84F near the wing tip. It was secured to the wing by bolts running through the wing spars. The jaw mechanism itself extended well forward of the leading edge of the wing.

Two hinged arms were mounted one above the other at the front end of the capture mechanism. During cruise flight the arms pointed straight ahead. They were enclosed in a streamlined clamshell housing that was painted yellow. A vertical, steel pin was located between the arms of the capture mechanism. Before approaching the wing tip of the GRB-36F, the arms of the capture mechanism spread to form a vertical fork that guided the pin to the receiver plate. The receiver plate on the GRB-36F had a semicircular recess in its rear edge to receive the coupling pin.

The width of the recess in the receiver plate was much smaller than the height of the open jaw capture mechanism. This reflected the fact that it was much easier to control the lateral placement of the Thunderflash with respect to the wing tip of the B-36 than it was to control the relative height of the two airplanes.

There was no latch to hold the pin in place after contact was achieved. The pilot of the Thunderflash held the coupling mechanism in place by using the thrust of its jet engine. He released the RF-84F from the mechanism simply by throttling back.

### Phase I Flight Tests

Doc Witchell piloted the GRB-36F, 49-2707 on a flight on December 12, 1954in preparation for the Phase I Tom-Tom flight tests.

The first phase of Tom-Tom flight tests was begun in early 1955. Tests were conducted to evaluate the handling of the parasite fighter as it flew in close proximity to the bomber's wing tip and the feasibility of capturing the towing arm in the wing tip jaw of the fighter.

Beryl A. Erickson and Raymond Fitzgerald piloted the test airplanes. They switched off duties. The two pilots each gained experience flying both the mothership and the parasite. Other GRB-36F crewmembers included Convair employees R. E. Fisher, B. White, and J. L. Blackman.

On the morning of a mission, the GRB-36F and the RF-84F took off from the Convair plant at Fort Worth and fly to the test operations area southwest

*RF-84F 51-1848 approached the left wing tip of the GRB-36F during the initial phase of Project Tom-Tom flight tests. (LMTAS Photo 26-1817)*

of the city. A Convair 240 airliner carrying a camera crew accompanied the Tom-Tom test airplanes on their flights.

The pilot of the GRB-36F flew straight and level at the test altitude. The coupling mechanism operator in the mothership extended the scissors mechanism and locked it in the fully extended position. The pilot of RF-84F 51-1848 maneuvered to a position behind and to the left of the GRB-36F. He opened the arms of the jaw capture mechanism.

The Thunderflash pilot moved in slowly until the jaw capture mechanism on his wing lined up with the scissors mechanism. He advanced the throttle slightly and contacted the receiver plate with one of the forward arms of the capture mechanism. The arm guided the capture mechanism pin into the recess on the receiver plate.

The pilot of the RF-84F used the thrust of the jet engine to hold the pin in place. He simply throttled back slightly to pull the capture mechanism pin out of the recess on the receiver plate and move away from the wing tip of the GRB-36F.

Erickson and Fitzgerald became reasonably adept at the hook-on maneuver. In light of their success, it was decided to proceed to the second phase of testing. There is no evidence that RF-84F 51-1848 engaged in Project Tom-Tom after the initial boilerplate tests.

#### *Full-scale structural wing tip coupling*

On August 17, 1955, additional funds were allocated to the Tom-Tom project for the installation and testing of the full scale coupling mechanism on

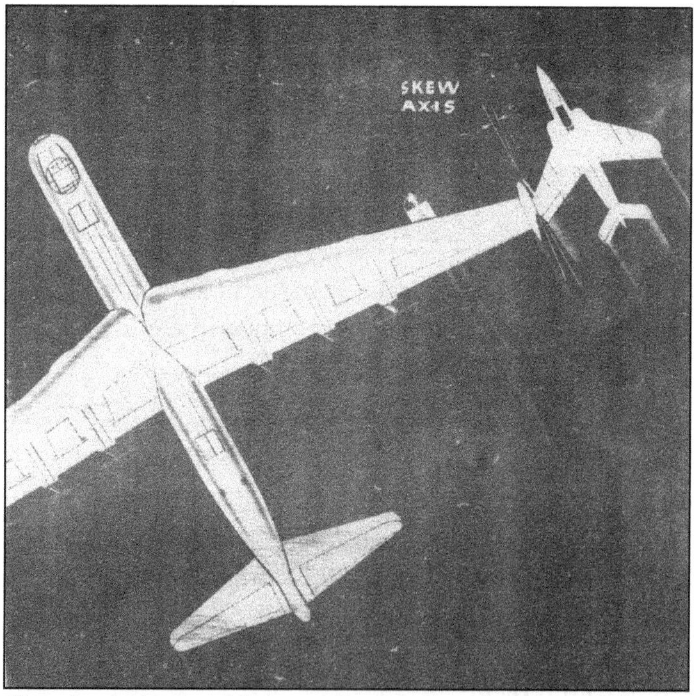


*The skewed axis of the coupling mechanism caused the RF-84F to pitch down as it rolled up and pitch up as it rolled down. (Tom-Tom Development Engineering Inspection Safety: Convair FZM-36-1191 via Bud Anderson)*

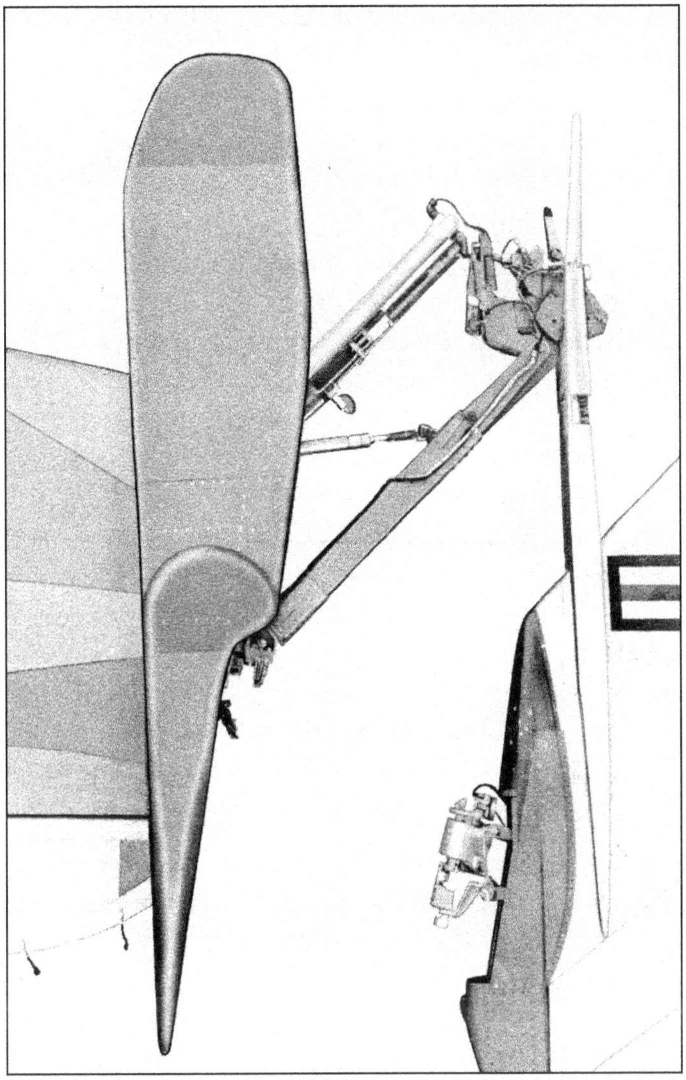

*One-point attachment. (FZM-36-1191 via Bud Anderson)*

both wing tips of the GRB-36F.

In late 1955, RF-84F 51-1849 was equipped with a fully functional, coupling mechanism on its left wing tip. The arms of the capture mechanism were redesigned. Rather than housing tubular members inside streamlined fairings, the forward end of the capture mechanism split

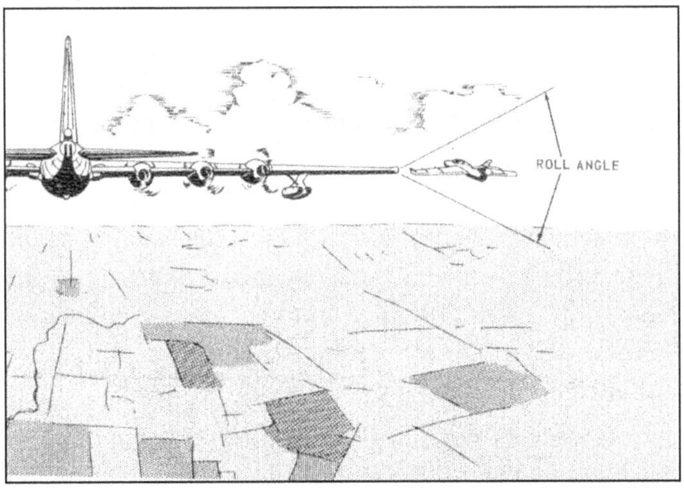


*A jettison mechanism was designed to release the RF-84F from the wing tip of the GRB-36F if the roll angle exceded preset limits. (FZM-36-1191 via Bud Anderson)*

*Two-point attachment. (FZM-36-1191 via Bud Anderson)*

*Only the wing tip coupling mechanism on the right wing tip of the GRB -36F was tested. (FZM-36-1191 via Bud Anderson)*

*A wing tip coupling mechanism was installed on the left wing of RF-84F 51-1849. (FZM-36-1191 via Bud Anderson)*

into a pair of streamlined arms. The arms were attached with explosive bolts to jettison the parasite in the event of an emergency.

The coupling mechanism incorporated a pitch lock near the trailing edge of the wing tip. During cruise flight, the pitch lock was housed in a well on the wing tip. It was extended from the well to engage a matching receiver on the wing tip of the GRB-36F. The pitch lock was pivoted to allow the RF-84F to rotate around the coupling mechanism.

Revised wing tip coupling mechanisms were installed on both wing tips of the GRB-36F for the next phase of Tom-Tom tests. Erickson stated that the full-scale coupling mechanism on the left wing tip was never completed, due to the difficulty of coupling with the left wing tip of the JRB-36F. No efforts were made to couple to the left wing tip in the second phase of testing.

The structure of the revised mechanism was much stronger than the earlier test rig. The skin of each wing tip of the GRB-36F was stripped off for a few feet inboard of the tip. The coupling mechanism was attached to the exposed wing spars.

The receiver plate was equipped with a latch to secure the pin of the capture mechanism on the Thunderflash. The scissors unit was hydraulically operated with electrical controls. It was possible to retract and extend the mechanism while the RF-84F was latched to the receiver.

The receiver for the pitch lock on the wing of the RF-84F was located to the rear of the scissors unit. While the Thunderflash was being towed from the receiver plate it was free to assume whatever pitch attitude was appropriate for its weight and the current flight conditions. As the wing tip of the Thunderflash was pulled toward the wing tip of the GRB-36F, the pitch lock engaged the receiver, locking the RF-84F at that pitch angle. The Thunderflash was then free to roll around the wing tip coupling, but it was fixed in pitch and yaw.

A substantial aerodynamic fairing was installed on the wing tip to house the pitch lock receiver and the retracted scissors unit.

The wing tip coupling mechanism incorporated a skewed hinge line like that tested in the NACA Langley wind tunnel tests. The forward attachment point was located outboard of the rear pitch lock. The aerodynamic effects of the skewed hinge line restored the RF-84F to the proper attitude when it rolled away from the desired position.

The wing tip coupling mechanisms were equipped with automatic mechanical jettison devices that would release the RF-84F if it rolled beyond preset limits.

The aerodynamic fairings that enclosed the towing arms of the GRB-36F were painted blue. The scissors units were painted red-orange. The jaw capture mechanisms and the rear pitch locks of the RF-84Fs were painted yellow. By painting the components of the airplanes different colors, traces of paint that transferred from one to the other made it possible to tell where they came into contact with each other.

The nose turret of the GRB-36F was removed and a large air data sensor boom was installed in its place. The boom and the fairing over the space formerly occupied by the turret were painted red-orange. An air data sensor boom was installed on the nose of each Thunderflash to place the sensors as far from the disturbed air around the wing tip of the GRB-36F as possible.

#### *November 1955, JRF-84F*

In November 1955 the designations of the Thunderflashes used for the Tom-Tom Project were changed to JRF-84F to indicate that they had been substantially modified for the tests, but that the modifications could be reversed to return them to standard RF-84F configuration. The designation of 49-2707 was changed from GRB-36F to JRB-36F.

#### *Ground test rig*

Ground tests of the wing tip coupling mechanism were performed using a gantry to lift the RF-84F high enough off the ground to hook on to the wing of the GRB-36F. A four-legged gantry with an overhead crane was constructed at the Convair plant.

To test the coupling mechanism, the GRB-36F was towed into place with its wing tip under the edge of the gantry. The RF-84F was hoisted off the ground to the level of the wing tip of the GRB-36F. The movement of the crane simulated the approach of the Thunderflash to the wing tip coupling mechanism. The yawing of the Thunderflash was controlled by a man standing to the rear of the gantry, holding on to cables that connected to each side of the tail of the RF-84F through pulleys mounted on the legs of the gantry.

Observers paid close attention to the function of the latch in the receiver plate. Then the scissors unit was retracted to pull the Thunderflash against the wing tip. The observers noted the behavior of the pitch lock

A wing tip mechanism was installed on the left wing tip of the GRB-36F, but it was not tested.. (LMTAS Photo 26-2294)

RF-84F 51-1849 in the ground test rig. Note the man standing below the Thunderflash controlling its yaw angle with cables connected to its empennage.. (LMTAS Photo 26-2152)

as it engaged the receptacle in the wing tip of the GRB-36F.

#### *March 21 - 22, 1956, Development Engineering Safety Inspection*

Convair hosted an inspection of the test airplanes at their Fort Worth plant on March 21 and 22, 1956 for the purpose of evaluating the safety of the wing tip coupling operation. On the first day of the conference, the attendees were shown movies of the RF-84F flying in the proximity of the wing tip of the JRB-36F and the scale-model wind tunnel tests that had been conducted at the Langley Aeronautical Laboratory. They examined the test airplanes. Convair technicians demonstrated the functions of the coupling mechanisms.

The RF-84F was mounted in the ground test rig for the second day of the conference. The attendees were treated to a demonstration of the entire coupling and uncoupling process.[xii]

Beryl Erickson and Raymond Fitzgerald prepared for their upcoming wing tip coupling flights by visiting the Beechcraft plant in Wichita, Kansas to fly the Long Tom XL-23C with its free-floating, wing tip fuel tanks.

#### *April 13-14, 1956 First Full-Scale Tom-tom test flight*

Full-Scale Tom-Tom test flights were initiated on April 13, 1956. Tom-Tom flight tests were conducted in a sparsely populated area to the southwest of Fort Worth. Thee JRB-36F carried as many as nine crew members during the test series. All flights included the two German engineers.

Beryl Erickson flew RF-84F 51-1849, with the coupling mechanism mounted on its left wing tip for the first time on April 13, 1956. He evaluated the affect that the wing tip coupling mechanism had on the flight behavior of the RF-84F in solo flight. On April 14 Erickson flew 51-1849 in close proximity to the modified right wing tip of the JRB-36F. This and all

*A technician directs the crane operator as he guides the jaw capture mechanism toward the receiver plate in the ground test rig. (LMTAS Photo 26-2160)*

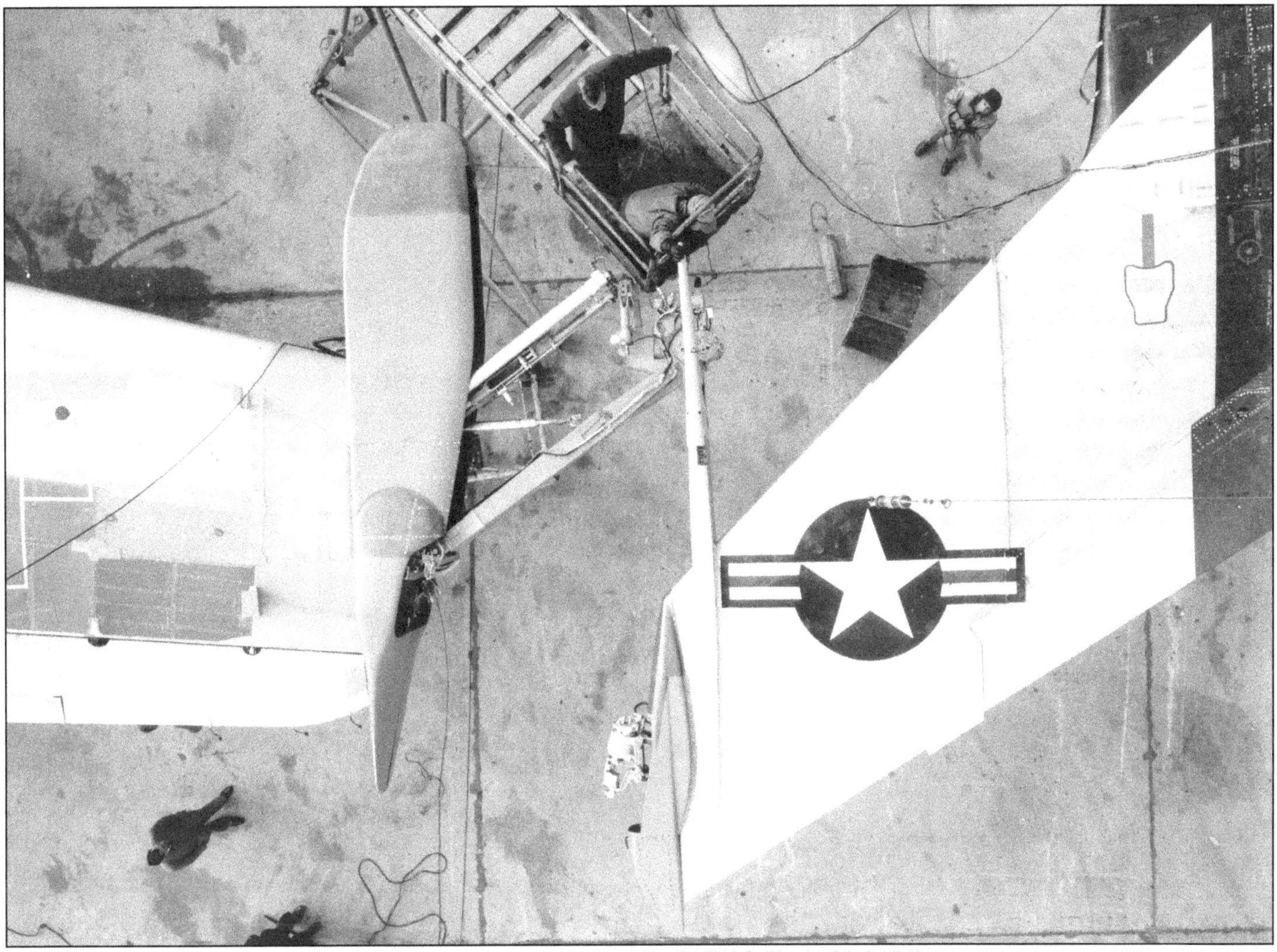

*RF-84F 51-1849 in the one-point position in the ground test rig. (LMTAS Photo 26-2154)*

Project Tom-Tom flights were flown with no external tanks or stores.

### *April 24, 1956 First Hook Up*

The first hookup of the revised wing tip coupling hardware was accomplished during a 1 hour and 30 minute long flight on April 24, 1956. A. S. "Doc" Witchell piloted the JRB-36F on this flight. Witchell was chief pilot of the Convair flight department. He had been the co-pilot on the first two flights of the all-jet YB-60 version of the B-36 on April 18 and 25, 1952. He regularly flew GRB-36D FICON carrier airplanes in 1954 and 1955. He had been the pilot of the first flight of the NB-36H nuclear reactor carrying version of the B-36 on July 20, 1955 and many of its subsequent flights. He made some of the last FICON retrievals in an RF-84K at Fort Worth in February 1956.

Between April 24 and June 8, eleven composite flight tests were made by the JRB-36F. Over the course of those flights, Erickson evaluated the process of hooking on to the wing tip of the JRB-36F and the behavior of the RF-84F in coupled flight.

On each test flight, Doc Witchell or Raymond Fitzgerald stabilized the JRB-36F in straight and level flight at the test altitude. The coupling mechanism operator manipulated the hydraulic controls to extend the scissors unit on the right wing tip. Erickson opened the jaw capture mechanism and triggered the extension of the pitch lock from its well.

The approach to the receiver plate was similar to the tests already conducted. Erickson stabilized his RF-84F momentarily at a position behind and to the right of the wing tip of the JRB-36F. He applied the rudder gently to line up the capture mechanism with the receiver plate and then advanced the throttle. The arms of the jaw capture mechanism guided the coupling pin into the latch of the receiver plate. The latch locked automatically as it engaged the pin, securing the Thunderflash to the receiver. This condition was referred to as "single-point attachment".

Erickson leveled the RF-84F wings relative to the wing of the JRB-36F and adjusted the trim of his elevator to reduce the force he applied to the control stick to zero. When the Thunderflash was properly trimmed,

*The scissors mechanism is retracted to engage the pitch lock in two-point attachment in the ground test rig. (LMTAS Photo 26-2166)*

Erickson radioed to the coupling mechanism operator to inform him that he was ready for the retraction of the scissors unit.

The coupling mechanism operator toggled the switch that controlled the retraction of the scissors unit. As the unit retracted, the pitch lock of the RF-84F engaged its receiver on the JRB-36F. This condition was called "two-point attachment". The movement of the Thunderflash was restricted to rolling around the hinge line of the coupling mechanism.

The skew angle of the wing tip coupling produced a stabilizing effect on the Thunderflash. When it was displaced from the wings-level position by gusts or by maneuvers of the JRB-36F, it stabilized at the proper position after just a few small oscillations. Erickson was able to fly "hands-off" for extended periods of time.

When Erickson was ready to conclude each test, he called to the coupling mechanism operator to extend the scissors unit. When the unit was fully extended, he throttled up the jet of his Thunderflash to relieve the drag on the latch mechanism. At his signal, the coupling operator released the latch and Erickson's RF-84F dropped back from the coupling.

Before landing, Erickson closed the arms of the jaw capture mechanism and retracted the pitch lock into its well. The coupling operator aboard the JRB-36F retracted the scissors unit into the fairing on the wing tip.

Pilots of the RF-84F had to learn to coordinate control inputs and throttle settings to engage the coupling mechanism. If the JRB-36F rolled to the left, its wing tip would rise. If the pilot of the Thunderflash just pulled back on the control stick, his airplane would slow down and he would fall behind the wing tip of the carrier airplane. He had to advance the throttle to climb up to the wing tip of the JRB-36F without falling behind.

The RF-84F pilot approached the wing tip from the rear. He stabilized his airplane with its wing tip a few feet behind the mechanism on the wing tip of the JRB-36F. He advanced the throttle to increase his speed just a few miles per hour to push the probe into the receiver. The Thieblot Company did not specify the maximum approach speed.

Once the probe was secured in the receiver, the Thunderflash pilot immediately reduced the throttle to avoid yawing forward around the wing tip of the JRB-36F. The RF-84F was free to rotate in pitch and yaw. Its pilot had to constantly adjust the throttle and elevator as the wing tip of the JRB-36F moved up and down in turns. Project pilots found that maintaining position in this configuration was no problem in smooth air, even while the pilot of the JRB-36F made turns of up to 40 degrees of bank angle in both directions.

Flight in the one-point position was particularly hazardous. If the RF-84F lost engine power at that point, the airplane would yaw away from the wing tip and roll downward. It would rapidly increase the drag on the wing tip of the JRB-36F, leading to a loss of control. Such an incident would result in the loss of both airplanes and their crews.

The scissors unit pulled the RF-84F snugly up against the wing tip, engaging the rear latch of the mechanism. With the aft latch secured, the RF-84F was limited to rolling up and down around the wing tip of the carrier airplane. The skew angle of the coupling insured that the RF-84F would align itself with the wing of the JRB-36F. The rear latch could be adjusted up and down to compensate for changes in the weight or load of the RF-84F.

On the fourth or fifth flight, Erickson shut down the engine for about fifteen minutes. He restarted the engine and prepared to separate from the JRB-36F. There was an electrical power failure to the wing tip coupling and jettison mechanism. Erickson was "marooned" on the wing tip. It was the one time during the project that he was really scared.

If the power could not be restored, there was a plan to dump the airplanes in the Gulf of Mexico. The crew of the JRB-36F would bail out. Erickson would eject from the Thunderflash and then the pilot of the JRB-36F would aim the airplanes out to sea and bail out.

An engineer aboard the JRB-36F restored electrical power to the wing tip coupling mechanism after one hour and ten minutes. Erickson restarted the engine of the Thunderflash a second time and separated from the wing tip feeling greatly relieved.

Doc Witchell piloted the JRB-36F on Tom-Tom flight tests on May 11, May 25, and June 6. He made a 1-hour 30-minute flight check of the JRB-36F on June 7.

Witchell piloted the JRB-36F for 4 hours and 10 minutes during his last Tom-Tom flight on June 8.[xiii] Erickson hooked his Thunderflash to the wing tip of the JRB-36F twice. Each time the scissors unit was retracted to draw the Thunderflash against the wing tip

*The Thunderflash approaches the right wing tip of the GRB-36F during full-scale Project Tom-Tom flight tests. (LMTAS Photo 90-3082)*

and seat the pitch lock. Erickson flew in the "two-point" position for 23 minutes during the first coupling and 45 minutes during the second.

Tom-Tom flights were temporarily suspended after the June 8 flight.

#### *August 31, 1956 Resumption of Flight Tests*

Tom-Tom test flights resumed on August 31, 1956. Six flights were flown between August 31 and September 26. Raymond Fitzgerald and Beryl Erickson piloted the test airplanes. They switched off so that each gained experience flying both airplanes. Although they became reasonably adept at hooking on to the wing tip of the JRB-36F, it usually required more than one attempt to capture the receiver in the jaw on the wing tip of the RF-84F.

A primary objective of the test flights was to evaluate the stability of the skewed wing tip coupling and the feasibility of flying long endurance missions while attached to the wing tip. The RF-84F was towed for as long as 1 hour and 30 minutes. The jet engine of the Thunderflash was shut down during long endurance cruise flights.

#### *September 26, 1956 Final Tom-Tom Mission*

Beryl Erickson encountered a highly undesirable flight condition during the eighteenth Tom-Tom flight on September 26, 1956. Erickson and Fitzgerald were aware that they were probably making the last flight of Project Tom-Tom. There was strong opposition to the project from the Pentagon.

The RF-84F Thunderflash was 1,500 pounds heavier than it had been on previous Tom-Tom flights. Doc Witchell flew the JRB-36F faster than before to increase the dynamic pressure on the airplanes during the test by approximately 15%.

Erickson rendezvoused with the JRB-36F ten minutes after take-off. The test aircraft were approximately forty miles southwest of Carswell AFB when Erickson coupled his airplane to the wing tip of the JRB-36F. On his first contact of the flight, Erickson decided to kick the rudder as he approached the wing tip of the JRB-36F to jam the probe into the latch crooked. The probe hit the latch with considerable lateral motion rather than straight forward. The latch engaged at an excessive yaw angle and the Thunderflash "diverged rapidly".

The right wing of the RF-84F dropped 15 degrees, then up 20 degrees in a diverging oscillation. Erickson directed his vision forward as the Thunderflash flapped up and down. He did not change the throttle setting. On the third "upfly" the RF-84F tore the scissors unit off of the wing tip of the JRB-36F.

Some of the wing tip coupling mechanism of the JRB-36F remained attached to the wing tip of the RF-84F. The increased drag was equivalent to four square feet of additional flat plate area. Erickson regained control of the Thunderflash and climbed back up to the JB-36F.

The Thunderflash started rolling up and down around the coupling mechanism. With each oscillation, the amplitude of the flapping increased until the latch mechanism at the end of the scissors unit was overloaded. The latch mechanism was torn from wing of the JRB-36F. The RF-84F veered away from the JRB-36F with the latch still gripped in the jaws of the capture mechanism on the wing of the Thunderflash.

Witchell and Erickson had no trouble flying the damaged airplanes back to the Convair plant in formation. Erickson landed his RF-84F ahead of the JRB-36F.[xiv]

The two German engineers examined the damage to the wing tips of the airplanes. They had the damaged components packed up and shipped to Wright field for analysis. They departed from Fort Worth immediately afterward. There was little in the way of subsequent investigation of the incident.[xv]

The following day, September 27, Convair recommended to the Air Force that the Tom-Tom project should be terminated. The Air Force agreed to terminate the project on November 20, 1956.

Convair technicians carefully documented the damage to the aircraft with photographs. The damage to the RF-84F was limited to some denting of the outboard surface of the jaw capture mechanism arms. The attachment points of the receiver latch absorbed most of the damage to the JRB-36F.

### Conclusions

By the end of September 1956 Fitzgerald and Erickson had accomplished approximately 50 couplings on 18 flights.[xvi]

The termination of the Tom-Tom project marked the end of Air Force attempts to launch and retrieve manned parasite airplanes from bombers.

*The damaged RF-84F 51-1849 following the final flight of Project Tom-Tom on September 26, 1956. (LMTAS Photo 26-2297)*

*The damaged RF-84F 511849 following the final flight of Project Tom-Tom on September 26, 1956. (LMTAS Photo 26-2300)*

Only one RF-84F hooked on to the JRB-36F at any time.

Convair's report about the Tom-Tom project concluded that the design concept and associated mechanisms were well thought out and engineered, and the operational feasibility was within operational program objectives. Fitzgerald and Erickson stated the opinion that the handling qualities of the Tom-Tom RF-84F were "tentatively acceptable".[xvii]

Autopilots of that time were not sufficiently precise to control the JRB-36F during the engagement of the wing tip mechanisms. The JRB-36F had to be flown with a steadiness that demanded the reflexes of a human pilot.

Operational fighter pilots did not like the project. Fighter pilots were fatigued by the long hours spent riding attached to the bombers.

Although flight tests had shown that the Tom-Tom concept was workable in a controlled situation, Erickson confided to this author that the program was "too far out" for operational use. There were concerns that a pilot might be too fatigued or injured to hook back onto the wing tip of the carrier airplane at the end of his mission. Any operational mission would have required the RF-84F pilot to remain cooped up inside the cockpit for much longer than anyone could endure.

In any event, an airplane was entering service with the Central Intelligence Agency that would be able to perform many of the missions envisioned for the Tom-Tom system without resorting to the complex operations of composite aircraft, the Lockheed U-2.

#### *Disposition of Test Aircraft*

Following the conclusion of the Tom-Tom program, the JRF-84F Thunderflashes had the special equipment related to the tests removed. In July 1957 they were flown to the Ogden Materiel Area at Hill Air Force Base, Utah. Their designations reverted to RF-84F.

In February 1958 they were flown to the Mobile Air Materiel Area at Brookley Air Force Base, Alabama. They were dropped from the inventory of the United States Air Force and transferred to the Military Assistance Program for delivery to the Greek Air Force.

In June 1957 the JRB-36F was flown from Fort Worth to the Military Aircraft Storage and Disposition Center at Davis-Monthan Air Force Base, Arizona and placed into storage. It was not stored for long. In August 1957 the Air Material Command authorized the reclamation of the materials that it was built from. After

*The damaged jaw capture mechanism of RF-84F 51-1849 following the final flight of Project Tom-Tom. (LMTAS Photo 26-2295)*

*The pitchlock of RF-84F 51-1849 following the final flight of Project Tom-Tom. (LMTAS Photo 26-2296)*

*The damaged jaw capture mechanism of RF-84F 51-1849 following the final flight of Project Tom-Tom. (LMTAS Photo 26-2308)*

*The damaged jaw capture mechanism of RF-84F 51-1849 following the final flight of Project Tom-Tom. (LMTAS Photo 26-2308)*

*The damaged jaw capture mechanism of RF-84F 51-1849 following the final flight of Project Tom-Tom. (LMTAS Photo 26-2295)*

*The pitchlock of RF-84F 51-1849 following the final flight of Project Tom-Tom. (LMTAS Photo 26-2296)*

The damaged jaw capture mechanism of RF-84F 51-1849 following the final flight of Project Tom-Tom. (LMTAS Photo 26-2308)

The damaged jaw capture mechanism of RF-84F 51-1849 following the final flight of Project Tom-Tom. (LMTAS Photo 26-2308)

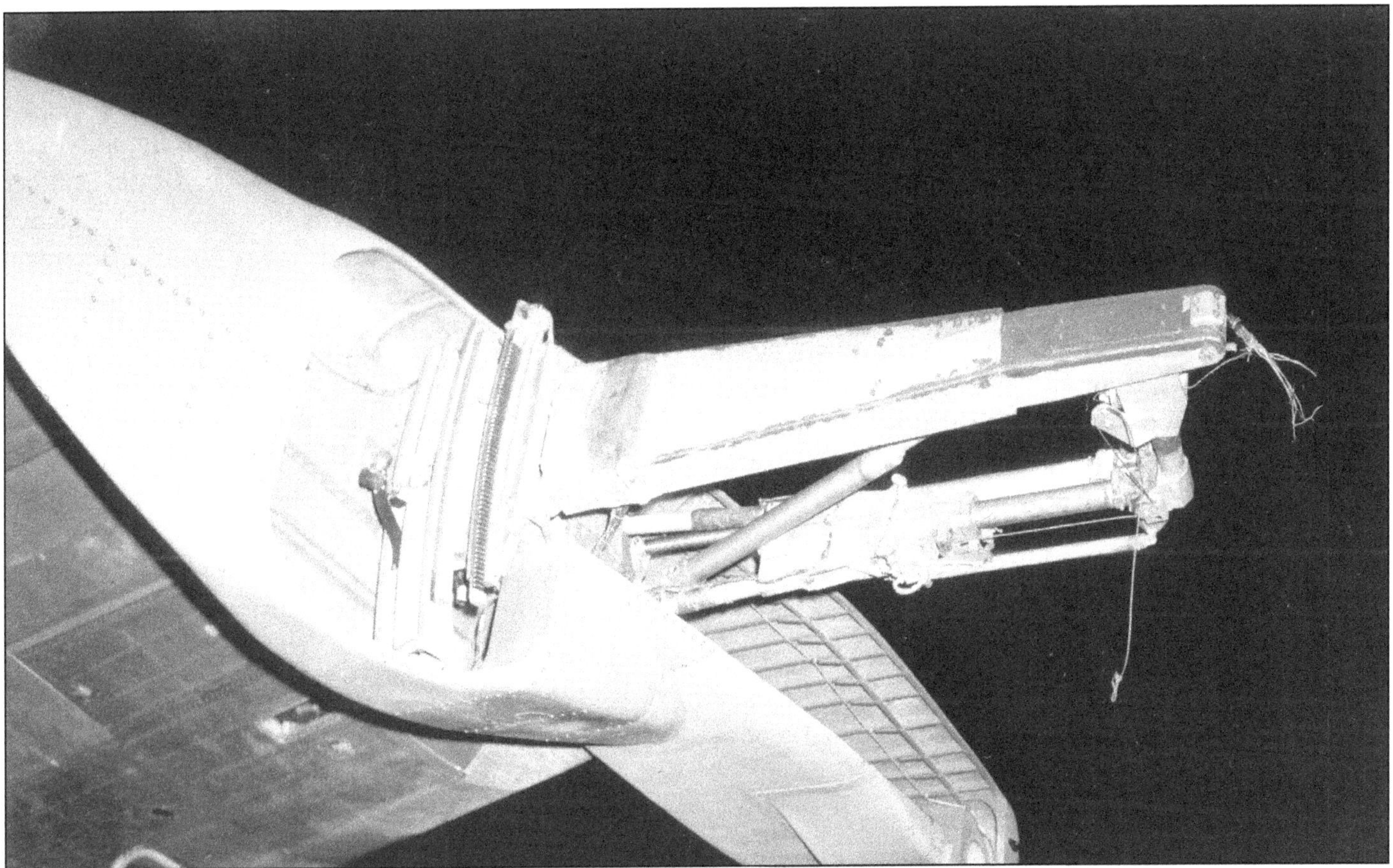

*The damaged scissors mechanism of GRB-36F 49-2707 following the final flight of Project Tom-Tom. (LMTAS Photo 26-2310)*

*The damaged scissors mechanism of GRB-36F 49-2707 following the final flight of Project Tom-Tom. (LMTAS Photo 26-2325)*

its instruments and engines were removed, 49-2707 was towed to the contractor's area on the south side of the base. It was cut into large chunks with a giant blade dropped from a crane. Its pieces were pushed into the blast furnaces and melted into ingots.

### Weapons system WS-110A

The free-floating wing tip fuel tank concept found its ultimate expression in the early phases of Weapon System WS-110A, which led to the development of the XB-70A Valkyrie. Both North American and Boeing generated proposals for large supersonic bombers with free-floating wing tip fuel tanks. The composite airplanes would cruise at subsonic velocity on the fuel contained in the wing tip tanks. When that fuel was depleted, the tanks would be jettisoned. The remaining component of the composite airplane would accelerate to supersonic speed for the dash over the target.

Boeing's entry was 156 feet long. The composite airplane had a wingspan of 232 feet. The manned central component had a wingspan of 70 feet.

It is claimed that, when General Curtis LeMay, the Commander of the Strategic Air Command, was shown these proposals, he said, "This isn't an airplane! It's a God damn three-ship formation!"

The eventual development of the compression lift theory by North American made it possible to design a bomber that could cruise at Mach 3 the entire distance to and from the target with resorting to free-floating wing tip fuel tanks.

### References


[i]NACA RM L9I07a: Flight Tests of a Model having Self-Supporting Fuel-Carrying Panels Hinged to the Wing Tips; Shanks, Robert E. and Grana, David C.; Langley Aeronautical Laboratory; November 1949

[ii]Research and Development Information Report 51S-36820; Engineering Division, Air Materiel Command; February 28, 1951

[iii]NACA RM L51A12: Free-Flight-Tunnel Investigation of the Dynamic Lateral Stability and Control Characteristics of a tip-to-tip Bomber-Fighter Coupled Airplane Configuration; Bennett, Charles V. and Cadman, Robert B.; Langley Aeronautical Laboratory; April 4, 1951

[iv]NACA RM L51E17: Free-Flight-Tunnel Investigation of the Dynamic Lateral Stability and Control Characteristics of a High-Aspect-Ratio Bomber Model with Self-Supporting Free-Floating Fuel Tanks Attached; Bennett, Charles V. and Cadman, Robert B.; Langley Aeronautical Laboratory; August 7, 1951

[v]NACA RM L52E08: Free-Flight-Tunnel Investigation of the Dynamic Lateral Stability and Control Characteristics of a High-Aspect-Ratio Bomber Model with a Sweptback-Wing Fighter Model Attached to each Wingtip; Bennett, Charles V. and Boisseau, Peter C.; Langley Aeronautical Laboratory; September 23, 1952

[vi]NACA RM SL55J24: Simulated Flight Investigation of a Scaled-Speed Elastic Swept-Wing Bomber and Fighter Models Coupled Wing Tip to Wing Tip; Thompson, Robert F.; Langley Aeronautical Laboratory; February 14, 1956

[vii]NACA RM SL56A25b: Flutter Tests of a 1/25-scale Model of the B-36J/RF-84F Tip-coupled Airplane Configuration in the Langley 19-foot Pressure Tunnel; Neely, Robert H.; Langley Aeronautical Laboratory; March 15, 1956

[viii]Trip Report: Flight Evaluation of Floating Panels on "Long Tom" Project; J. R. Baker, Jr.; September 26, 1957

[ix]Interviews: Beryl Erickson by author

[x]Calculated Lift Distributions of a Consolidated Vultee B-36 and two Boeing B-47 Airplanes Coupled at the Wing Tips; NACA Research Memorandum RM L50I26; November 30, 1950.

[xi]Disposition Form: Development Engineering Inspection of Project Tom-Tom; Lt. Colonel Leon S. Jablecki, Aeronautics and Propulsion Division; November 12, 1954

[xii]Tom-Tom Development Engineering Safety Inspection; Convair FZM-36-1191; March 21-22, 1956

[xiii]"Doc" Witchell flight log

Republic F-84 (Swept Wing Variants); Kevin Keaveney; Aerofax Minigraph 15; Aerofax, Inc.; 1987; Page 9.

[xiv]Interviews: Beryl Erickson by author

Republic F-84 (Swept Wing Variants); Kevin Keaveney; Aerofax Minigraph 15; Aerofax, Inc.; 1987; Page 7

[xv]Republic F-84 (Swept Wing Variants); Kevin Keaveney; Aerofax Minigraph 15; Aerofax, Inc.; 1987; Page 9

www.ingramcontent.com/pod-product-compliance
Lightning Source LLC
LaVergne TN
LVHW082355200926
848481LV00012B/3570
* 9 7 8 0 5 7 8 0 3 1 8 6 6 *